HOMEBUILDING

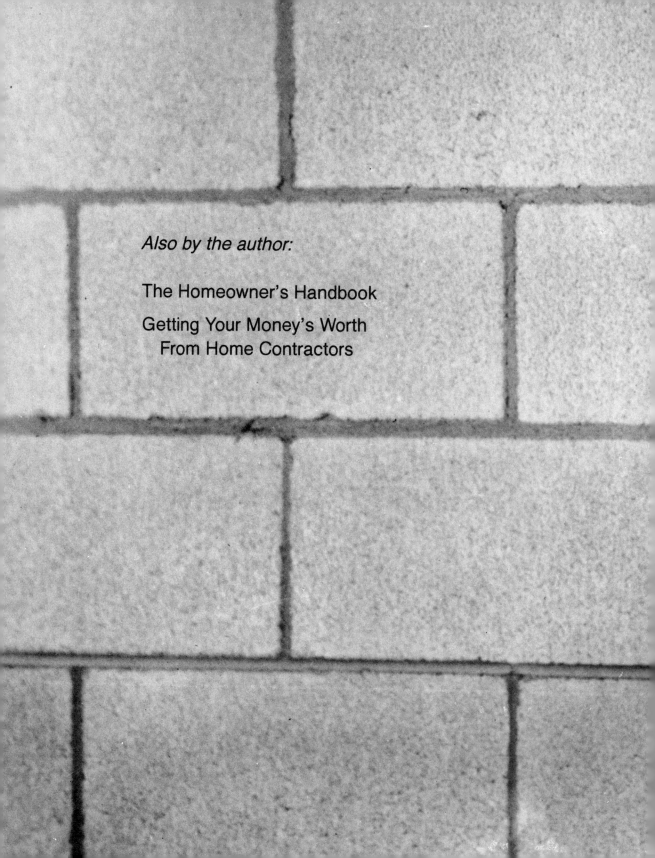

Also by the author:

The Homeowner's Handbook

Getting Your Money's Worth From Home Contractors

HOMEBUILDING

MIKE McCLINTOCK

A Comprehensive Guide to Footings, Foundations, and Framing Systems for Solid, Trouble-free Houses

Charles Scribner's Sons   New York

for Tek

The author appreciates the assistance and materials provided by the following manufacturers, organizations, and government agencies: American Institute of Timber Construction; American Plywood Association; Federal Trade Commission; National Association of Home Builders Research Foundation, Inc.; National Forest Products Association; National Ready-Mixed Concrete Association; Portland Cement Association; Southern Forest Products Association; Trus Joist Corporation; U.S. Department of Agriculture Forest Service; U.S. Department of Commerce, National Oceanic and Atmospheric Administration, Environmental Data and Information Service; U.S. Department of Housing and Urban Development; Western Wood Products Association.

Copyright © 1982 Mike McClintock

Library of Congress Cataloging in Publication Data

McClintock, Michael, 1945–
   Homebuilding: a comprehensive guide to footings, foundations, and framing systems for solid, trouble-free houses.

   Includes index.
   1. House construction.   I. Title.
TH4811.M335   690'.837   81-23184
ISBN 0-684-17445-6   AACR2

This book published simultaneously in the United States of America and in Canada—
Copyright under the Berne Convention.

All rights reserved.
No part of this book may be reproduced in any form without the permission of Charles Scribner's Sons.

1 3 5 7 9 11 13 15 17 19   y/c   20 18 16 14 12 10 8 6 4 2

Printed in the United States of America.

CONTENTS

About *Homebuilding* vii

Introduction ix

PART 1
FOOTINGS AND FOUNDATIONS 1

1. **The Nature of Footings and Foundations** 3

    Structural Stresses 4
    Planning Ahead 6
    Cost, Efficiency, and Safety 7

2. **Site Selection and Building Orientation** 11

    Points of View 14
    Aesthetics 14
    Time and Money 16
    Neighborhoods 19
    Site Analysis 22
    Orientation 24
    Legalities 27

3. **Masonry Tools and Equipment** 28

    Layout Tools 30
    Handling Tools 34
    Placing Tools 35
    Cutting Tools 36

4. **Footing Construction and Concrete Fundamentals** 38

    Excavation 39
    Soil Preparation 46
    Form Size and Design 48
    Pouring Concrete Footings 52
    Curing and Final Concrete Strength 59

5. **Foundation Construction and Block Fundamentals** 61

    Block Fundamentals 63
    Concrete Block Construction 67
    Poured Concrete Construction 82
    Wood Foundation Alternatives 91
    Weatherizing Foundation Walls 94

6. Slab Construction *101*

Concrete Mixes *103*
Vapor Barriers and Insulation *104*
Reinforcement and Control Joints *105*
Slab Finishing *108*

PART 2
FRAMING *111*

7. The Nature of Framing *113*

Conflicts of Design and Technology *113*
The Man-made Fortress *116*
Breaking Design Barriers *118*

8. Framing Tools and Equipment *121*

Length of Service *123*
Quality Cutting Edges *124*
Hardening *124*
Methods of Manufacture *125*
Layout Tools *126*
Construction Tools *129*
Construction Equipment *136*
Table Saws and Radial Arm Saws *139*

9. Properties and Design Values of Framing Members *141*

Framing Timber Sizes *145*
Lumber Design Values *146*
Interpreting a Lumber Design Value Table *148*
Interpreting Span Tables *153*
Conversion for True Rafter Size *154*
Design Guidelines *165*
Framing Alternatives *165*

10. Floor and Wall Framing *171*

Framing Systems *172*
Nails, Lag Bolts, and Frame Hardware *186*
Frame Construction Details *191*

11. Roof Framing *203*

Roof Designs *203*
Roof Construction Details *209*
Rafter Placement *214*
Trusses *217*

Foundation and Framing Information Sources *220*
Index *221*

ABOUT *HOMEBUILDING*

This book is not written only for professionals, although it contains a lot of technical information. And it is not written exclusively for avid do-it-yourselfers, although it contains a lot of practical advice. It does not assume that you have completed a carpentry apprenticeship or spent your summers working for a mason. And it does not assume that you can or want to build everything yourself, that you know how to use all the tools and materials, or even that you would recognize them in the lumberyard.

Homebuilding is written for everyone who is interested in the place where they live. It is a meticulously thorough guide that includes everything you need to know about designing and building every structural component of a house, from the smallest piece of 2 × 4 blocking to the largest new home, from the footings to the roof ridge, from the most insignificant nailer to the most crucial beam.

Included are comprehensive and practical explanations in layman's language of all the materials, tools, and building procedures used in footing, foundation, and framing work on residential structures. The material characteristics, structural relationships, ratios, formulas, and building techniques for a wide variety of designs are presented in manageable form and in the proper sequence of construction.

In this way, knowledgeable readers can pinpoint information on a particular subject, and those without building experience can use the book as an overall guide to do the work themselves or to evaluate the proposals and work of professionals they hire.

This is not a textbook. It does not talk about the imperfect process of construction solely in technical terms, because houses are not created in test tubes by scientists in white coats. They are built under widely varying field conditions by people with diverse skills and interests.

Framing timbers may be heavy with moisture or a little warped, or dumped off the delivery truck into a mud puddle. You may not be able to get a concrete truck in to the job site. The excavation may fill with water from the only five straight days of rain in years. The young building inspector may be out to prove his detailed knowledge of the most far-fetched building codes.

Given these possibilities and the diverse interests, talents, and personalities of the people involved, it is unlikely that any job, even a small one, will run according to plan. And any plan for construction based on regimented procedures and a dogmatic, right way to do everything will prove too rigid for diverse and changing field conditions.

Homebuilding takes this into account. It includes all the options and explains how to cope with unforeseen material and design changes and the inevitable structural monkey wrenches.

It covers the possibilities and parameters of material selection, substitutions, deliveries, storage, job relations, job sequencing—all the peripheral elements that can make or break even the smallest structural improvement. And it includes the professional tricks of the trade that are regularly used to bridge the gap between theoretical situations and the real thing.

It is a book you can study to learn the theories of building. It is also a book you can consult to get timely, practical guidance on every aspect of structural residential construction.

INTRODUCTION

From the street both houses look alike. They are both large, white clapboard colonials with two stories of double-hung windows, comfortable front porches, and brick chimneys capped over steeply sloping roofs. The house on the left was built fifty years ago; the house on the right, only one year ago.

The older house has required two new layers of asphalt shingles over the original roof, a new furnace, a new water heater, and many other replacements. The furnace and most of the appliances in the newer house are still under warranty and working efficiently.

Close up, you can see caulking around the windows and doors of the fifty-year-old structure, although it shows no signs of cracking or separating from the joints between siding and trim. Unexpectedly, the same joints on the new house have also required caulking, but here it has worked loose, leaving many of the joints open to the elements.

Behind the lush shrubbery of the older house, smooth stucco, tinted green with mildew, covers the stone foundation. On the newer house small shrubs permit enough ventilation and sunlight to eliminate mildew formations. But at two corners of the foundation, there is cracking along the hidden joints of the new concrete block and chunks of stucco have broken off.

The siding on the fifty-year-old house has been coated by seven thick and slightly uneven layers of paint, but the top layer shows no signs of peeling or flaking. On the one-year-old house some nailheads are visible through the single, even coat, and at the corners the paint is disrupted and several clapboards are splitting where the nails are driven through. The more you look, the more confusing it gets. The new house has so many obvious problems, the old house so few.

If, to unravel this paradox, you called in an experienced residential building contractor, architect, or engineer, you could probably get a very accurate picture of what to expect when you looked inside—even details about building materials, the condition of walls and trim, and more.

How is this possible? Simply because the lack of deterioration on the older house indicates that it is strong, tied together securely as a single structure, and stable on its site. By comparison, the obvious problems on the newer house indicate that it is not very strong, that one side of the house is structurally disconnected from the other, and that the house is moving too much.

To experts, the pattern of deterioration on the new house may indicate thin concrete mixes, inadequate soil consolidation, lack of cross bridging, and other failings. To even the most inexperienced homeowners this pattern only reinforces a suspicion many of them already had—houses aren't built as well as they used to be.

To be fair, new is sometimes better. One advantage you can expect in a new house is efficient and maintenance-free operation of plumbing, heating, and electrical systems. Compared to those installed fifty or even twenty years ago, you should get more usable heat from the fuel you buy (even though its cost has soared) as well as a longer lifespan for the mechanical equipment. Old water heaters, for instance, were lined with metal and easily accumulated energy-wasting layers of corrosive mineral deposits. Modern heaters are glass lined, which virtually eliminates this problem.

You might expect, though, that over a fifty-year period construction techniques would have progressed as dramatically as the efficiency of mechanical systems. Nevertheless, the front door on the new house closes with a hollow rattle, unlike the reassuringly solid thud of the door on the old house.

Why are so many new houses structurally inferior to older ones? The answer lies in the circumstances and attitudes of three interested parties: home builders, home sellers, and home buyers.

Homebuilding is a competitive business. It suffers dramatic highs and lows according to the vagaries of interest rates and other economic conditions. It is seasonal and risky. These pressures make many builders exceptionally thrifty with time and materials, and they engender a construction process where minimal amounts of marginally adequate materials are assembled as quickly as possible. As reasonable as this may sound (at least from the builder's point of view), the minimal approach to building tends to produce minimally adequate and, all too frequently, inferior houses.

Some of the problems created by corner-cutting construction are isolated and cosmetic, like a hastily applied, two-coat taping job that doesn't fill up wallboard joints and starts cracking after a few seasons. But most of the problems occur where their effects can be extensive and damaging—in the basic structure of

INTRODUCTION

the house: the foundation and framing.

Building is a logical, sequential process. A mistake in the final stages of construction is relatively easy to correct because it is accessible. But an uncorrected mistake in the early stages is buried, difficult to correct later, and has a snowballing effect that makes each successive piece of construction more difficult.

A miscalculation in the foundation and framing changes the positive process of adding layers of modular materials to a negative one of accommodating them to previous errors. For example, if the first floor framing has not been squared or well braced during construction, the careful, 16-inch, center-to-center stud layout may break down. Eight-foot pieces of Sheetrock miss the corner studs, and on the exterior wall, siding panels come up short. What can be done? Add an extra stud, fill in the gap with a thin strip of Sheetrock, and hope the taping job will cover it? Once the structural mistake has been made, there is no cosmetic remedy that provides an effective, long-term solution.

Enter the second interested party, home sellers. They pay surprisingly little attention to structural soundness and concentrate instead on finishing touches: the wallpaper, the dining room chandelier, the brick veneer fireplace, the oversize tub in the master bathroom.

Naturally, this tendency rubs off on the building contractor or developer. The real estate agent isn't pressing for kiln-dried frames for the fixed glass, so why bother. If the demand for quality and costly materials and procedures isn't there, why supply them. But multitone door bells, cork wallcoverings, and built-in vacuum systems? That's another matter.

These finishing touches are high on the builder's and the real estate agent's list of goodies. They dress up model homes and provide fodder for copywriters of sales brochures and newspaper ads. They give you something to point to, to talk about, to admire, and, most importantly, to remember when it comes time to choose between one new development and another.

It's difficult to drum up this kind of interest in full-perimeter foundation drains, or termite shields, or five-ply, exterior-grade subflooring nailed off to architectural specifications—as difficult as it would be for an airline to advertise the advantage that their planes have strong wings.

Most homeowners take quality construction for granted or simply accept what they get, good or bad, without question. That's a mistake. Of course, an agent's sales spiel does not make allowances for a question like, "Yes, but when my kids wrestle in the second floor bedroom, will the living room ceiling wind up in my lap?" It may be embarrassing to ask, but the enormous cost of a house should be justified with more than cosmetic goodies and a slick sales pitch.

It is ironic that serious questions about the most important parts of construction are the ones asked the least or not at all.

They are neglected because crucial structural components of a house are not common knowledge and not visible. This is a deadly combination that permits marginal and corner-cutting construction to pass unnoticed behind a veil of memorable and salable extras.

The third interested party, the home buyers, must accept some responsibility for buying structurally inferior houses largely on the strength of cosmetic features. Consumers are comfortable with the rule, "What you see is what you get," and while it may work in the supermarket it does not work in the housing market.

Even experienced homeowners find it difficult to evaluate structural quality. While years of homeowning may turn many of them into proficient painters, paperhangers, upholsterers, or even cabinetmakers, most lack the opportunity to discover the fine points and sound techniques of structural masonry and carpentry work. The skin is generally better understood and easier to appreciate than the skeleton.

However, this is not because structural work is more technical than finishing work. Matching complex wallpaper patterns around a bow window can be just as taxing as laying out a stud wall. And it is not because structural work requires more skill than finishing work. Its gross nature commonly permits wider tolerances for fit, and requires less hand tool proficiency than cabinetwork. It is strictly a matter of opportunity and experience.

Practical work in the field is helpful in learning any skill, but it is not the only way to become knowledgeable. That's why this book covers many types of construction in detail, including the variations in materials used, tools required, and techniques employed to build solid, long-lasting, trouble-free homes. It purposefully peels away those attractive, salable extras to expose the crucial structural components of residential housing.

Is this knowledge valuable? If you build your own house, your own addition, or even your own set of porch steps, it's essential. If you hire a professional to do all or part of the work it's invaluable. For once, you'll be in a position to judge accurately the true quality of materials and labor you are paying for and to determine if the job is planned and executed correctly.

Understanding how important the structure is, what its elements are, and how they should be put together can save you time and money during construction and over your full term of ownership. When the skeleton is sound the materials that transform a bare frame into a home are easier to apply, require less maintenance, look better, and last longer.

HOMEBUILDING

PART 1
FOOTINGS AND FOUNDATIONS

1

THE NATURE OF FOOTINGS AND FOUNDATIONS

The most obvious example of a footing is the hinged piece of anatomy at the end of your leg. Simply, it provides a base, a kind of platform, on which the weight of your body rests. In natural structures, great amounts of material and sophisticated physiology are expended to provide footings. Plants and trees develop extensive systems of roots, which, in addition to balancing them, allow them to shift and follow the light. The tip of the iceberg remains visible only because of the great, buoyant mass beneath the water that keeps it upright.

The structural necessity of a footing system is easily demonstrated. Try balancing a sharpened pencil on its point. Then try balancing a new, unsharpened pencil (this is possible on a true surface, although tricky), and finally a pencil stuck into a footing such as a block of clay.

People and buildings respond to footing support the same way. This is why it is difficult for a ballerina to dance on her toes, even with special toe shoes. The concentrated load of body weight requires extra strength at the point of support and extra balance to keep the load upright. Sure, it's a simple principle of physics, and an obvious structural relationship that is grasped quickly by preschoolers playing with blocks. Surprisingly, engineers sometimes overlook its importance.

By calculating the center of gravity, it would be easy to cast a metal statue of a ballerina raised up on her toes without special braces or reinforcement. And if human beings were stationary creatures who conducted all their business telepathically, a pad or platform footing would be unnecessary. The leg bone could be locked at right angles to a single, stubby foot bone the way an anchor bolt is fixed in a block of concrete.

But human beings do move and dance and carry groceries and run for buses and shovel snow. They are constantly performing operations that change the relationship between body and footing. A redistribution of body weight (bending over, for instance) alters this relationship by placing the load away from a vertical line above the center of gravity. Adding a live load (carrying heavy objects or pushing up with your arms to unstick a balky window, for example) also causes a change.

Man-made structures undergo similar stresses, and the alterations that occur in both types of structures can be accounted for with a thrust line, a modern-sounding term that was first used in the late eighteenth century by Charles Augustin de Coulomb. This French scientist, who gave up careers as a superintendent of waters and fountains and as a military engineer, is remembered more for his work on electricity and magnetism than for his innovative explanations of structural elasticity. Coulomb used the idea of a thrust line to describe the path of stress through materials, which at the time was a completely novel concept. To understand this movement, it is necessary to make the distinction between two types of stress, compression and tension.

STRUCTURAL STRESSES

Compression is a pushing force and tension is a pulling force. Every object is stressed in compression simply by the action of gravity on its mass. Consequently, compression is an easily identifiable and understandable force that was accounted for in the design of man's earliest structures. But tension, the pulling force, which can act against the stress of gravity, was not seriously investigated until the nineteenth century.

Picture an inflated balloon resting on a table. If you push down on one side of the balloon the elastic material underneath your hand is compressed and takes up less space. At the same time the wall of the balloon opposite your hand starts to stretch. As the tension in its elastic surface increases it grows thinner and less opaque, while the compressed surface under your hand grows more dense and more opaque. Furthermore, by changing the position of your pushing hand, and, therefore, the direction of the compressing force, you can shift the compression and tension stresses to different areas of the balloon's surface.

Both these forces are present in all but the most perfectly balanced, perfectly vertical walls. Consider a brick wall, one foot wide, that is built to these impossibly perfect specifications. Brick, which has a compression resistance, or crushing strength, around 7,000 pounds per square inch, weighs approximately 150 pounds per cubic foot. So every foot of height added to the wall increases the compression load by 150 pounds. At a height of 10 feet, 1,500 pounds compress the bottom layer of bricks; at 100 feet compression is 15,000 pounds; and at 1,000 feet the compression load is 150,000 pounds.

Remember, though, that this load is applied on one square foot of brick. The load on a square inch (there are 144

The Nature of Footings and Foundations

square inches per square foot; therefore, 150,000 divided by 144) is just over 1,000 pounds, or only one-seventh of the force needed to crush the brick at the bottom of the wall. The 7,000-pound crushing strength of brick leaves six-sevenths of this strength intact, which is an extravagant safety margin.

The high compressive strength of masonry (stone is about the same as well-made brick) is largely responsible for the survival of ancient temples, palaces, and houses of the wealthy who had the time and resources to build with permanent materials. But given the safety margin on a 1,000-foot-high wall, you might expect to find many masonry buildings with vertical walls approaching this magnitude. However, the highest surviving ancient structure is Khufu's pyramid, the largest of the great pyramids at Gizeh. It is 479 feet high but 767 feet square at the base and certainly not a vertical wall.

The explanation is the force of tension. A strong breeze, mild vibration from a distant earthquake, or even a head-on impact from an ox-drawn cart all could have collapsed the exceptionally safe (from a compressive point of view) 1,000-foot-high wall. While a great amount of weight could have been added on top of the wall, only a small amount of sideways force would have been needed to create a fatal tension stress.

An oblique stress would increase the tension on the outer skin of the wall the same way tension stressed one side of the balloon. In masonry, a crack would appear in the mortar between bricks, then the crack would widen, and finally the bricks on top of this tension tear would tip over.

In the ideally vertical wall, the thrust line passes straight down its center. But the sideways pressure of wind, for instance, forces it off center. As the thrust line moves toward the outer skin of the wall (the outer third of the wall's thickness is a danger area), tension increases and produces cracks. When enough force is applied to move the thrust line outside the boundaries of the wall, the downward compression stress is replaced by lifting tension stress, and the wall topples.

You can easily simulate this displacement of the thrust line. When you stand upright the load of your body behaves like the load of the vertical wall, and is transferred straight down to your feet. The position of the thrust line creates a symmetrical distribution of weight around your center of gravity, and this makes it easy to remain upright. Standing with your arms somewhat away from your sides, grasp a relatively heavy object like a telephone in your right hand. As you pick up this extra weight you can feel muscles and tendons in your right ankle respond to the weight imbalance. You may also be able to sense a decrease in pressure, or the beginning of a tension stress, in the footing system muscles opposite the extra weight on your left side.

The marvelous thing about this process is its relegation to the unconscious mind. The number of mathematical proportions and constantly changing relationships of this sort that are calculated by the brain and executed by the body every day would keep several structural engineers pinned to a computer.

Equally staggering calculations may be necessary for the efficient and safe construction of buildings and particularly for massive structures like bridges and dams, where forgetting to carry the tens can have catastrophic consequences. One of the most memorable cases in point is Galloping Gertie, which was the frighteningly accurate nickname of the Tacoma (Washington) Narrows Bridge. Due to a severe miscalculation of stress (in this case, torsional stiffness), the bridge roadway was sent into an undulating movement by a paltry 42-mile-per-hour wind. The wave pattern in the roadway increased, causing excesses in compressive, tensile, and other stresses, until the span snapped and collapsed into the river below.

Stiffness in man-made structures, as opposed to resilience in natural structures, is not always desirable. In fact, one of the tallest structures in the world, a 2,000-foot television transmission tower near Fargo, North Dakota, is designed to sway over 13 feet in 120-mile-per-hour wind gusts. Better to sway than to snap. Even relatively tall office buildings containing brittle glass in fixed frames and other rigid materials can accommodate tension stress. The Empire State Building has been measured as swaying close to 2 feet at the top, which may be alarming but certainly not dangerous for the people inside as they watch the New York skyline move back and forth.

A capacity for resilience, for a design reserve that can withstand extraordinary compressive and tensile stresses, is a fairly recent development in buildings. And it is the chief difference between natural footing systems and construction footings.

PLANNING AHEAD

Natural footings are highly resilient. A tree bends in the wind without uprooting, your ankle twists and tenses as you bend without snapping, only because the connection between the load and the footing it rests on is flexible. When you stand, the muscles and tendons in your feet and ankles provide only enough counterforce to keep you upright. As you bend or redistribute your body load, this counterforce increases by increments proportional to that stress. In other words, the tension stress reserve is applied efficiently and only as it is needed.

By comparison, construction footings are static. The connections between a concrete footing and a masonry block wall are secured with mortar and steel rods. In a typical house there is little provision for bending when a freak storm deposits 3 feet of water-heavy snow on the roof. It is that once-in-a-lifetime storm or tornado or undermining from flood water erosion that must be taken into account before the footing is built.

The United States Army Corps of Engineers, in a project to determine the design of a breakwater surrounding a nuclear power plant off the coast of New

Jersey, had to predict the worst storm that might occur in the area. To hazard a guess at possible storm damage to such a sensitive structure based on last week's forecast would be foolhardy. In fact, while the Corps of Engineers routinely uses 50- or 100-year time spans for these estimates, a 1,000-year frame was used for the nuclear project.

The worst storm in 1,000 years was characterized as a double hurricane, a condition caused when a passing storm is swept back to the site as a second hurricane is arriving. While the relatively shallow waters around the proposed plant were considered to be capable of supporting 31-foot waves, this one-in-a-thousand condition was predicted to push an additional 23 feet of water into the area—enough to support 48-foot waves.

The differences in construction materials, techniques, and costs between a breakwater with a one-in-fifty storm resistance and a one-in-a-thousand resistance are considerable. But a one-in-a-thousand safety margin offers a lot more security than one in fifty.

Certainly the planning and safety factors of major structures used by the public, or whose destruction would affect the public, are more critical than similar considerations on a house. But the principle behind these calculations is the same: Footing design must foresee all dead weight (building materials), and all live loads (people, wind, snow, etc.) that may possibly be carried.

The only caveat is the qualifier "possibly." Theoretically, at least, everything is possible. But everything is not practical, physically or financially. This leads to a pragmatic balancing act between a margin of safety that provides physical security and a cost of construction that is economically feasible.

In a sense, this stringent preplanning is also a part of natural footing design, the chief difference being the time allotted for thousands of years of natural selection, or natural genetic engineering. If people were suddenly mutated to creatures with one huge, fifty-pound hand, our present footing system would be inadequate. The time of evolution provides a cushion of trial and error and the possibility of adjusting components into proportional, integrated development that any builder or engineer would relish.

The development of man-made structures, however, is bound by time, and also limited by material availability, manpower, and money. And many consumers who have bought cars and swimming pools and houses can testify that a reasonably priced piece of engineering may not necessarily be reasonably safe.

COST, EFFICIENCY, AND SAFETY

There are few works big or small that this triangle of parameters does not affect. It includes pieces of work that don't cost so much that no one can pay for them, that don't hurt you of their own accord, and that perform the function for which they were designed with acceptable thoroughness.

Notice the order in which they are listed. Cost, the ever present bottom line, used so frequently to hogtie aesthetics, utility, and even safety, is the first star in the triangle. Of course any piece of engineering can be made safer if more time and money are spent furthering this particular interest. Beams can be larger, concrete more dense, quality control more selective—it's just a question of money. The case against the Ford Pinto gas tank, for instance, was alleged to be one of safety suffering because of a marketing plan. In other words, price is the predetermined package into which safety, along with efficiency, is stuffed regardless of fit.

Efficiency can also be improved if more time and money are devoted to its cause. More extensive consumer tests and field monitoring can be done with prototypes, which are revised and improved and retested until they provide high-quality service. Almost any physical by-product of the space program demonstrates this. Turn enough scientists loose with enough time and equipment, and they will explore every eventuality, every stress, every action and reaction, until a thorough, functional design, complete with safety valves and backup systems, is produced.

However, cost, predictably the most limiting factor in construction, has also been the positive catalyst for many engineering innovations. This is evident in the development of dam construction. The whole purpose of a dam is to withstand sideways pressure. If the force of water behind a dam was somehow applied directly downward, its walls could be vertical. And the high compressive strength of masonry would make the construction economical. But the sideways pressure of water against a dam creates a massive tension stress that requires buttressing, bracing, and stabilizing.

Excessive tension stress on the inside skin of a dam wall must be controlled to prevent cracking, which can allow water to enter the structure. On a house foundation, similar cracking may lead only to a wet basement due to slowly seeping groundwater. But on a dam, even though many contain interior drains to deal with seepage, water entering surface cracks exerts a tremendous lifting force (the force of tension) of five tons per square foot at a depth of 100 feet. Once the wall is weakened by tension cracking, this lifting pressure can open the cracks more and push the dam over.

To compensate for this pressure dam walls used to be built with increasing thickness toward the base. At the top of the dam there is little water to apply sideways pressure, and the thrust line will run down the middle of the wall. Toward the dam base, however, water volume and the resulting pressure is increased. This extra force would push the thrust line outside the boundaries of the wall if it were not thickened, a process that added considerably to the time and cost of construction.

On modern dams the resistance to tension stress that used to be accounted for by great thickness is now counteracted by less expensive steel reinforcement running from the top of the wall into the bedrock below the dam. While this pretensioned steel does not increase compressive strength, it dramatically increases tensile strength to prevent collapse by a staged process of cracking, water infiltration, and finally, hydraulic lifting.

A graphic example of this process can be seen in a movie called *The Dam Busters,* a slightly romanticized account of the destruction of the Mohne and Eder dams that overlooked the industrialized Ruhr Valley in Germany during World War II. The crux of the British attack was conceived by an English scientist named Barnes Wallis, whose plan to collapse the dams by crippling their footings was met with great skepticism.

To accomplish this, he developed a round bomb casing for high explosives to be released at very low altitude on the reservoir side of these dams. Due to the forward velocity of the aircraft, the bombs bounced along the water surface like stones that are skipped on smooth water. Wallis's calculations, and the points of release, were so precise that the bombs stopped skipping just as they approached the dam wall. There, instead of exploding outright, they slowly sank down to the dam footing and were triggered by a depth gauge. Accounts of this attack detail a time lapse between the explosions and the collapse, due to the staged destruction process of tension cracking, then water infiltration, and finally, an immense hydraulic lifting pressure that tipped the structures over.

This attack, and its success, demonstrates how crucial a footing system is. Shake it, and the structure is likely to collapse. In fact, Wallis's plan was only a modern extension of medieval siege warfare, which routinely included attempts to undermine enemy fortifications. This type of attack, and the devastation it wrought if successful, led to the development of moats around castles and fortresses, in part to make it difficult for attackers to reach the walls, but particularly to keep them from tunneling under them.

This work has been done throughout the history of siege warfare by sappers. In World War I, tunnels not unlike shallow mine shafts were dug to points directly under enemy trenches and filled with explosives. In Erich Maria Remarque's *All Quiet on the Western Front,* there is a terrifying moment, as one front line company is relieved by another, when word is passed to the newcomers that the sounds and vibrations of digging from somewhere beneath their position have just stopped, signifying the placing of explosives.

In medieval siege warfare, sappers would tunnel under the footings of a fortress, propping up the great weight and thickness of the masonry walls with wooden beams as they dug. Then, when the footings were sufficiently undermined, these beams were set on fire. Although medieval sappers and their commanders did not have the benefit of Wallis's advanced mathematics and physics, they did have the sense to realize that the defensive force of a massive battlement could be turned against itself by undercutting the feet. This, after all, was only the structural equivalent of a common fighting technique called hamstringing, in which a well-armored, shield-carrying warrior was made defenseless by severing his Achilles tendon.

Through the different stages of building—using different materials, tools, and techniques, and working with varying degrees of understanding about how structures stand up—footings have been a crucial key to form and function. From the understandable and easily applied properties of gravity and compression to the recently explored effects of tension stress and structural elasticity, the final product has always depended on a solid base.

The paramount importance of this construction component to builders throughout history has made it possible for twentieth-century tourists to walk between the 2,400-year-old Doric columns of the Parthenon. It has made it possible to pass through the double arches of the 1,700-year-old Porta Maggiore, which also supports two aqueducts in what is left of a fortification around Rome consisting of 383 towers, over 7,000 lookout posts, and 13 monumental gateways.

It is even more significant that through the history of warfare the footing has been selected for specialized attack—from hamstringing to undermining medieval battlements, to tunneling explosives under enemy trenches, to tension cracking a dam footing to cause its collapse. The crucial target has always been the footing—only the method of attack has changed with the technology of the time.

Again, footing construction is not as critical in a single home as it is in massive, public-use structures. And the consequences of poor design and underbuilt construction are not likely to be catastrophic. But many homeowners are familiar with nailheads protruding from wallboard, with cracks in trim and walls around door openings, with window and door frames that work out of alignment, with block foundation walls that crack open and let groundwater into basements. In fact, inadequate structural design and construction is one of the reasons why plaster is no longer used on house walls. Current structures can't hold it in place securely enough to prevent tension cracking.

These and other signs of deterioration, which can frequently be traced to an inadequate footing system, may be mundane next to a collapsing bridge span. But while many of these conditions are only cosmetic, they are symptoms of a basic structural problem whose effects can change the character of a home, detract from its appearance, decrease its value, and dramatically alter the amount of time, money, and effort you must devote to the place where you live.

2

SITE SELECTION AND BUILDING ORIENTATION

If you travel an hour by car in any direction from a large city you will have a difficult time finding a high-quality building site. Almost all of them have been used up. These building sites, which are the products of wind, rain, soil composition, erosion, and other factors, take time to develop. And as products of natural forces they ensure a harmonious coexistence between the site and the surrounding terrain.

Of course, you can create good building sites by carting in clean fill, grading it carefully to provide drainage that may be assisted with pipe- or tile-lined trenches, and by landscaping to prevent erosion and provide efficient thermal control. However, these man-made landscapes are mini-environments that may work well as independent islands, but are likely to conflict with the larger environment surrounding them.

A common complaint of new homeowners is a wet basement. The soil composition and slope around one house, even a row of houses, may be adequate. But if the development site has interrupted a naturally constructed runoff channel, or if most of the ground cover and trees that held the soil in place have been removed, the natural forces of the surrounding environment will overwhelm the seemingly adequate defenses of each building.

The available selection of good sites is small. And the complete reshaping of development landscapes may disguise the few spots where basements are likely to stay dry year after year. But it wasn't always this hard to find a good place to build a house. Homesteaders crossing the country in covered wagons were tempted to stop and settle down when they came upon a highly suitable rangeland or valley, a maxi-environment in which they lived and worked, and from which they derived food, fuel, and aesthetic satisfaction. Food came from the ground after

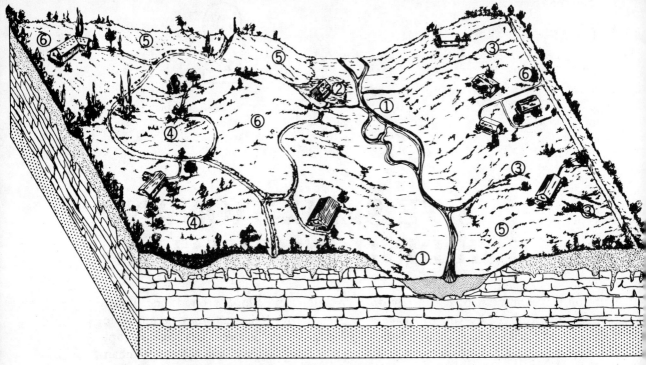

Figure 1. Building sites: 1) Flood plain—subject to flooding during heavy storms; 2) Alluvial plain—soil formed as water erodes material from watershed above; can be damaged by flash floods; 3) Upland waterway—water flowing from higher surrounding ground concentrates here; 4) Depression—collects water; soils remain wet and spongy for long periods; 5) Hillside—soils may be shallow and subject to slippage and erosion; 6) Rise—deep, well-drained soil on ridges and sloping hillsides; generally the easiest site to manage.

seasons of planting and cultivating, not from the local supermarket. Fuel came from the forest after days of cutting and hauling, not from a gas pipe, oil truck, or electrical cable.

Ah, but the struggle: rising before sunrise, eating dinner after sunset, plowing under a hot sun, hunting and trapping in the freezing snow, hauling hewed beams out of the forest, carrying lake water to the kitchen. How hard it must have been to spend most of your waking hours busy with the basics of survival. But this way of life was cut short by the closing of the West, and the growth of towns and cities. In fact, the conventional paths of progress to increased specialization and technological assistance have all but precluded the possibility of finding a homesite on the scale of a four-hundred-acre valley.

Many of the benefits of living in an environment with enough space and natural wealth to sustain a family and a productive life have been exchanged for conveniences. Now, in all but the few pockets of truly rural land that remain undevel-

Site Selection and Building Orientation

oped, building sites are not self-sufficient. Most modern building sites are clustered to take advantage of communal structures like libraries and hospitals and to make communal services like police protection economically feasible. By banding together, citizens can defray the cost of these services: i.e., you may use the full length of Main Street but pay only for a small, theoretically proportional, share of its initial cost and ongoing maintenance.

The homesteader did without these conveniences that have gradually come to be considered essential. In lieu of them, the homesteader could have surveyed the valley and selected a site on a natural rise, bracketed by runoff channels feeding a clearwater stream. The site would be protected on the northern exposure by stands of conifers, and alternately shaded in summer and exposed to the sun in winter by deciduous trees on the southern exposure—a process recently rediscovered under the title of passive solar heating.

However, in most suburban developments profit comes first. Considerations of natural drainage, soil composition, natural thermal control, efficient solar orientation, and countless other important building considerations affecting safety, durability, cost, maintenance, resale value, and overall quality, all play second fiddle to the powerful bottom line—profit. Trees, natural high spots, and area drainage channels are bulldozed into a flat tract. Access roads are cut and proportioned to conform to complex averaging factors that enable a developer to down-zone each building plot as a payback for access road surface area. This new, man-made environment is likely to be dressed up with sod farm grass, and nursery grown bushes, and this makes it difficult to pick out the good sites and pass up the bad ones.

The effects of poor site selection or inadequate site preparation are also noticeable in urban environments, particularly

Figure 2. The simplicity and efficiency of an ideal site, on a natural rise, open to sun and ventilation, protected from northern winds.

with condominium construction that approaches an assembly-line process. While this type of housing has emerged as the most profitable for builders—it's in demand more than single-family housing and is more affordable—units are being turned out wherever a market can be found and as quickly as possible.

Before proceeding to detailed considerations of site selection, keep in mind that the analogy of a homesteader in an idyllic valley is not intended to characterize the search for a good building site as a utopian pilgrimage. In a sense, the site is a subfoundation and subfooting to the structural elements built on and in the ground. Give it the same attention as the rest of the house, because its characteristics can be transmitted through every phase of construction. Picking the wrong site can adversely affect resale value, the amount of time and money you spend on maintenance, and other completely practical matters.

POINTS OF VIEW

Every site is different, even in developments where houses are made from the same blueprints and lined up like toy soldiers on terrain that has been homogenized to make every front lawn and backyard nearly identical. The first house in the row will catch northern winter winds head on. The second house will be a bit warmer, but the fireplace will not draw so well because winds around the first house cause turbulence around the next chimney in line. Differences may be subtle. In fact, while floor plans may have options for a bay window or extra closet space, most development homes are sold with the same warranty, the same basic building materials, and the same set of recommendations in the sales pitch.

That's one side of the contract. On the other are home buyers, and they have very little in common. They may be working with an architect on a one-of-a-kind site, shopping for handyman specials that need extensive rebuilding, or making the round of model homes in one school district as opposed to another. And while home buyers share the need for shelter, safety, security, and other matters, each one has a different view and a different interpretation of these characteristics and orders them in a different priority.

AESTHETICS

First impressions are lasting. If you go to look at a building site, and you have to walk down a freshly cut, muddy access road to a stand of trees with a lot number nailed to a stake, you assume you are buying a house that will be built on wooded land. This may seem like a highly logical assumption, but unless the builder has marked the trees, and specified in a contract that the finished site will include all those birches and pines you are looking at, this is probably the last you will see of them.

A case in point: Signs appear announcing a new development, something like

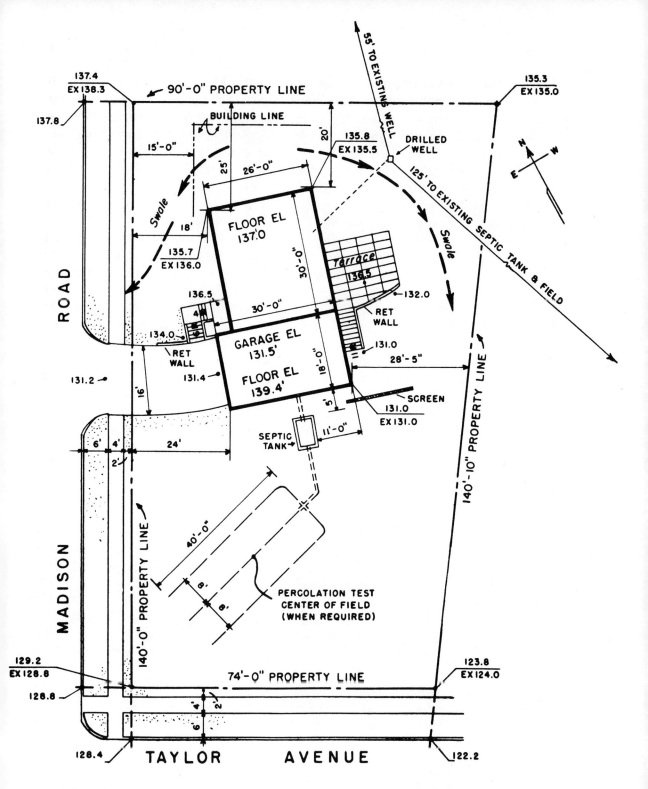

Figure 3. The complexity of a suburban lot, requiring soil tests, area drains, landscaping, and piping to create a site with characteristics that require maintenance.

Tall Pines or Thousand Pines, alluding to the dense, mature, evergreen-covered terrain; then the pines are cut and removed; then access roads are cut and the terrain is leveled; then houses are built; then, when multicolor flags are strung around the model home to attract buyers, about a hundred two-foot-high baby pine trees arrive from the nursery and are planted in a line behind the development. Adding insult to injury, about 20 percent of the young trees do not survive the first year and turn brown, but of course they are now the new owner's responsibility.

This particular inexplicable process of denuding and relandscaping occurred in an expensive metropolitan New York suburb, but it happens all over the country, regardless of housing prices. So when your decision about one site over another, or one development over another, is influenced by the lay of the land, the amount and type of landscaping, the view, and other aesthetic considerations, you must be sure that the terrain you look at from a muddy access road will resemble the same view from the freshly paved road.

Once you have determined what the finished site will look like, aesthetic considerations become highly personal. Some people like woods and hills, some like lawns and flatlands. But whatever your preference, you should give some consideration to the comparative aesthetic attraction of your homesite to the other sites in the area. It may be difficult to put yourself in the position of a home seller while you are busy with the details of home buying and building, but purely aesthetic considerations will affect resale value. For instance, statistics show that in suburban areas, corner houses are burglarized more often than similar homes on culs-de-sac. If you buy the piece of land on the corner and over the years your home is robbed at a rate consistent with current crime statistics, it may be difficult to sell.

Certainly this is a consideration that few people would take into account. It is, however, only one of many unrecognized site characteristics that can affect your home life and your home investment. Generally, you should be aware of how your site differs from neighboring lots. If there is a power transmission line in the area, can it be seen only from your living room window, or is that the only living room in the development where the landscape is more natural than technological? Will the morning sun warm your kitchen, or will this pleasant and fuel-saving feature occur only on the homes across the street that have an opposite orientation?

It is difficult to make evaluations about the practical side of personal, aesthetic impressions in one visit to the site. If possible, make several inspections, in the morning, in the evening, after a heavy rain, noting where and when the sun hits the building site and where water is likely to collect and run off.

TIME AND MONEY

For many, these are the most important considerations: initial cost, taxes and interest, ongoing maintenance and repairs, appreciation of land value, increased equity, strong resale value—all legitimate concerns about a large sum of money figured over the full term of ownership.

When you include the cost of construction (or the purchase price of an existing home) with the land value, the total price should not be more than 2½ times your annual income, while the regular monthly payments should not be more than 25 percent of your monthly income. These are general guidelines. There is no rule of thumb for the proportional cost of construction to land. Obviously, a small, modest house on a prime two-acre site will tally on one side of the scale while an elaborate house occupying most of a small site will produce an opposite proportion.

Looking past initial expenditures, you should tabulate the total costs (materials and labor) of improving the site. Most new home buyers spend money to complete jobs that should have been done by the builder. A revealing report, released jointly by the Federal Trade Commission and the Department of Housing and Urban Development, showed that an average new home has built-in defects that cost about a thousand dollars to repair. The leading category of complaints was inside the house, but the second most common group of problems included poor yard drainage and deteriorating driveways or exterior concrete work, both of which are directly related to proper site selection and preparation. Nearly 17 percent of the new homeowners surveyed across the country reported yard drainage problems, while over 15 percent indicated problems with driveways. Significantly, of all the categories surveyed, from interior finishing and painting to roofing and mechanical work, the one reported most frequently as not completed by the builder was grading and landscaping.

This indicates that builders have the same tendency as home buyers, i.e., to concentrate their attention on the house, not on the land under and around it. And while you may be able to rectify area drainage problems with rock and tile-lined area drains, or halt topsoil erosion by planting ground cover, it is very difficult to beef up an underbuilt footing and foundation system. Digging with a backhoe close against an existing house is a marginal proposition; and digging, partially undermining, temporarily supporting, and adding a new concrete pour to a deteriorating foundation is a time-consuming, expensive process.

If in doubt, consult an architect (one with extensive, practical field experience who will not be limited to solving this kind of nitty-gritty problem antiseptically and theoretically on a drawing board), or a structural engineer. On new building sites, be sure to establish the details of soil composition, the local water table, whether the ground contains a nearly impenetrable layer of hardpan or claypan, and the provisions for foundation and area drainage. Find out what trees, if any, will be removed; ask if the stumps will be left, pulled in one piece during excavation (thereby presenting a disposal problem), or reduced to chips by a stump removal machine (which is another extra expense).

In addition to the improvements you or the builder must make to the site, you must consider the services provided to the site that alter its value and, therefore, rate of assessment. For tax purposes the site is listed as "improved" as soon as these services are provided. They include paved access roads, which may be maintained by the town or in some developments collectively by the inhabitants; sidewalks; poles for electrical and phone lines; and water and sewer lines.

If the building site is well off the road, the local utility company may only install the first pole at its own expense, leaving responsibility for the rest of the run to you. This additional expense may crop up with sewer and water connections as well. In this regard, it is important to determine the trends of development in your area by visiting the local planning board or by tracking down a copy of what is euphemistically called the master plan for your township or county.

A considerable amount of planning, construction time, and money will be required for site improvement if water lines are not available (you will have to dig your own well). If, in very rural sites, gas lines are not available and there is no clear access for an oil truck, you may have to use bottled gas. And if no sewers are available, that means a complete private septic system.

There is also a catch-22 here, an expensive one, particularly if the site is on the edge of a growing residential community. You may go to the trouble and great expense of installing a septic tank, a leaching field, a drywell for clothes washer drainage, a 500-gallon oil tank (underground), and, after only a few years, be compelled to hook up to a new sewer system, and be economically forced to hook up to a new gas main. If town water is provided you can, of course, continue to use a private well. But sewers are likely to be provided as part of a regional project with heavy federal funding, and in the same way that you must pay taxes to support local schools (even when your children don't use them anymore), you must pick up your share of the sewer tab.

Other improvements can affect the time and money priority in subtle ways. For instance, a remote site with a long single-access driveway may be described as landlocked. If most of the property is bordered by private land instead of a public road (a small proportion of frontage), your real estate taxes may be substantially lower than those of other homes in the neighborhood with the same amount of property. A corner plot with twice the taxable frontage of typical homes in the area may bear a disproportionate share of local taxes.

Variations in site characteristics can cost time and money in two ways: first, as one-time outlays that are likely to pay for themselves in long-term, increased equity; and second, as ongoing expenses that are likely to increase right along with your equity. Frontage proportion, sewer connections, and other factors that are reflected in real estate tax rates must be examined over the expected term of ownership. For instance, if the site is near a pressurized hydrant your fire insurance rate is likely to be reduced. But if the site access road is too narrow for a fire truck—a culvert crossing may not be able to support the weight of the truck—the insurance rate may rise, year after year.

The conventional expectation of most new home buyers and builders is long term. But the typical house changes hands in less than seven years. That's why banks collect most of the interest during this period and use your payments to reduce the principal to a greater degree as the loan term progresses. Even at shorter terms of ownership, though, the site will require maintenance, another major consideration of the time and money point of view.

Maintenance may be divided into one-time versus ongoing outlays as well. Ongoing work qualifies as care and feeding, i.e., mowing grass, watering, pruning, fertilizing, the kinds of duties that most homeowners expect. Still, the type of site you select, and the way you prepare it, can alter the care and feeding schedule. For instance, the more land you grade flat and seed, the more time you will spend mowing. The more deciduous trees you leave over the lawn area, the more time you will spend raking.

On the other hand, one-time maintenance qualifies as repair and replacement work, which is a sinkhole for time and money. Carting in fresh topsoil to replace earth lost by water erosion or trying to replace trees that died because they were scarred by a bulldozer blade during excavation work may be investments you have to make, but they will never pay off. Spending money to regrade around the house perimeter to prevent drainage into the basement is not the kind of home improvement you can advertise. Buyers expect a dry basement. They expect the large shade tree in the front yard to be alive. And they expect the rich earth around the house to support ground cover, not to run in a muddy stream to the nearest storm drain every time it rains. Time and money considerations are important but not always the highest priority.

NEIGHBORHOODS

Everyone is affected by what their surroundings look like. That makes aesthetic considerations important. And everyone has financial limits. That makes time and money considerations important. But basic decisions about where to buy or build are most frequently determined by the character of the neighborhood.

This is particularly important to families, and they represent the single largest group of home consumers—about 40 percent of attached-house buyers, and nearly 60 percent of detached-house buyers. The point is, you want to build a home where you feel at home. It sounds simple, but neighborhoods are composed of many different factors.

Schools

Many families look for a building site based on school quality. The number of sessions, class size, distance from the site, facilities provided, and curriculum should

Figure 4. The Five Points intersection, New York City's Lower East Side, about 1829

all be considered. Also bear in mind that high-quality schools usually require high-level funding, which directly reflects on your real estate tax rate.

Stability

Even though the average term of home ownership is seven years, that is enough time for a neighborhood to change drastically. And although real estate is one of the few investments that has outstripped inflation in the last decade and even though vacancy rates are very low around most metropolitan areas, your equity and mobility can be undermined by buying and building in a destabilized community.

Zoning plays an important role in maintaining steady growth, or at least an equilibrium, in a community. If, for instance, land near your site is slated for down-zoning and concentrated condominium construction, area population can jump because this type of housing is most affordable to young families with children. This in turn may require increased taxes or bond issues to support additional school construction. Conversely, areas with regressive tax rates and little relief from commercial owners may create a stagnant housing market. After seven years of home maintenance and improvement you may have to slash the selling price well below assessed value because buyers will be hesitant to absorb your tax rate.

Security

Closely linked to stability, security is important for personal safety and for protecting your investment. For instance, even though most volunteer fire departments are exceptionally dedicated and efficient, in many suburban communities it

Site Selection and Building Orientation

will take them fifteen minutes or longer to respond with a minimal amount of equipment. Most private homes are not equipped with sprinklers and are constructed with wooden frames, unlike the aluminum or steel frames covered with noncombustible materials that are common in high-rise construction. This combination makes it likely that a furnace fire or major kitchen fire will severely damage the house before help arrives.

Police protection is important as well. Suburban burglaries are increasing at a time when police department budgets are being curtailed. It is not uncommon for a panel truck or van to pull up in front of a house, empty it, and depart in less than ten minutes. An increasing number of households have two job holders with regular schedules living in a home with minimal security protection. Without an active police department your possessions and insurance rates are in jeopardy.

Services

In addition to community hospitals, libraries, public parks, recreation programs, and retail facilities, all of which contribute to neighborhood quality, inquire about direct site services, including all utilities, garbage pickup, bulk removal to a town dump or landfill (this is crucial information that must be obtained before construction because building debris accumulates quickly), and adjacent road maintenance. In northern climates snow

Figure 5. State Street, New York City, 1864

removal, sanding, and salting can represent a surprisingly large proportion of local taxes.

The more information you acquire, the more likely you are to select the right site. And while it may be difficult to determine long-term neighborhood trends, there are national trends that affect almost all communities. As land, material, and labor costs increase it gets more difficult for young families to afford detached, suburban housing. The building and banking industries have tried to preserve the conventional dream of owning your own home, and mowing your own grass, using no-frills construction and a mystifying variety of flexible financing arrangements. But the inevitable trend is away from the single-family home and toward condominiums and attached houses. This trend is already well under way in many communities that, at first glance, seem to be stable.

In fact, communities can be stable to the point of stagnation. Restrictively zoned suburban communities are experiencing a rise in taxes and in the median age of homeowners. Schools built to accommodate their children have become part of a tax albatross. High mortgage rates discourage buyers. New construction and sales of existing homes decline. Slowly, the community develops concentrated, linear demographics with few high- or low-end homes, and becomes accessible to a smaller number of economically similar home buyers.

Economic fences can be difficult to break down. But as home buyers adjust their expectations toward smaller, energy-efficient homes on smaller plots, and as builders and planners work out the problems of multiple dwellings, even vertical, stagnating communities can be revitalized.

Finding a site with a predictable future is important, physically and financially. But whatever point of view carries the most weight with you, give some consideration to aesthetics, time and money, and neighborhood characteristics. Together, they exert powerful influences that shape the place where you live.

SITE ANALYSIS

There is a new trend in camping tents. Many are now made as self-supporting structures complete with waterproof floors, vents, and rain flaps that you can plunk down just about anywhere. But houses are heavy. If you plunk them down on a bog they will sink. If you build them in a natural swale they will flood. If you pour a foundation over inorganic silts or fine sands in a cold climate, severe frost action can tear the structural elements of the house apart in a single season.

To prevent these and other catastrophes, the physical properties of the building site must be thoroughly analyzed. Before a foundation is laid, particularly in a remote or one-of-a-kind site, you should gather information about the soil mechanics. Soil engineers, and some foundation engineers, will be able to specify allowable soil-bearing capacities. The usual

Site Selection and Building Orientation

procedure calls for an SPT (standard penetration test) hole to be drilled at each corner of the proposed building. From these samples the groundwater level, soil pH, and soil type can be determined. In special situations where local building codes and survey information are inadequate, laboratory tests can be performed to determine dry density, moisture content, permeability, and other properties.

As a general rule, the most secure anchor for a house foundation is bedrock. Forming against irregular rock contours can be time consuming, but with steel pins set into the rock as anchors, an exceptionally stable foundation can be built. This situation frequently requires a stepped foundation (raised and lowered to match the rock), and can present serious problems if a private septic system and leaching field are required.

Well-graded gravel, or gravel-sand mixture, with little or no fines (grit that passes through a $3/16$-inch square-mesh sieve) is a good foundation material with marginal frost action and excellent drainage. Silt-laden gravels and clay-gravel mixes are highly efficient as well but provide poor drainage.

Sandy soils (sand-silt and sand-clay mixtures) are a fair foundation material with only moderate frost action. Organic and inorganic silts and clays generally make poor foundation materials, have moderate to heavy frost action, and offer poor drainage. Highly organic soils like peat are not acceptable.

Soil identification can often be made without field investigations by tracking down old surveys or by obtaining geological reports and maps (available for most of the populated areas in the country) from the U. S. Geological Survey. Standardized methods of identifying soil are available in the Unified Soil Classification, adopted by the Corps of Engineers and the Bureau of Reclamation.

Foundations are not flexible. And in order for them to maintain their strength and hold the superstructure they support securely, they must be laid on soil that can withstand the structural weight with even, minimal settling. Almost all foundations settle a little. But hard, sound rock has an allowable soil-bearing pressure of 60 tons per square foot, while silt soil rates at only $1\frac{1}{2}$ tons per square foot.

Soil on the site may also be used as fill, to establish grades that slope away from the perimeter of the house, and to create flat lawn areas. Compacted fill up to a depth of three feet can be placed uniformly. Past this depth, uneven settling that leaves water-collecting low spots is more likely. Be extremely cautious about using fill (even if labeled clean fill) from other sites. It may come from roadwork where large amounts of broken concrete and asphalt are mixed in. Settling problems can also be caused by buried construction debris.

Drainage on the site must be analyzed as well. Water ranks as the most deteriorating element acting on houses. And although great pains may be taken to bring fresh water into the home for drinking, cooking, and washing, equal efforts must be made to keep groundwater out. On undisturbed sites you do not need elaborate tests to determine drainage patterns. Water flows downhill. Before it reaches the foundation it should drop into an area drain (a natural or man-made trench) that carries it away. Topsoil that is soft enough to poke a stick into and that rests on gravelly or sandy subsoil will hold water. Inorganic clays, hardpans, and claypans will act as spillways, bringing groundwater against the foundation. One possible result is a wet basement. But more serious, structural deterioration can be caused if the groundwater washes out the compacted soil under the footings.

ORIENTATION

Before any forms are built and any concrete is mixed you must decide how the structure will be oriented on the building site. Too often, this crucial decision is reduced to one very shaky convention—put the front door toward the street. Why? Because that is the way into the house? Not as a matter of course. Houses have more than one door, and the one used most frequently leads to the area inside the house that is the most active, for instance, a side or kitchen door. Many homeowners rarely use the elaborate front door that opens into a hallway. Instead, they may enter through the garage.

Houses are automatically placed with a street face orientation again and again, regardless of other important factors. However, building orientation should be the product of aesthetics (the position of the house on the site that is most attractive), physics (the relationship of the house to natural forces from sun, wind, light, water, and temperature), and social patterns as opposed to empty conventions.

Pure aesthetics (the look of evergreens reflected in a wall of sliding glass doors, the proportion of a roof against a ridge line) is a consideration that has no right or wrong conclusion. There are accepted principles of design, but if they dictate an orientation that you don't like, don't follow them. If you are working with an architect you can ask for a rendering of the house on its site before the plans are approved. If a model home has been built, you can photograph it and your particular site from different angles, have the site pictures enlarged, and then superimpose the house pictures against them. This inexpensive preview can be achieved by constructing a mask model (one without the details of windows and doors) to scale, and viewing it against the site blowup, as well.

Orientation according to social use is another factor with many variables. The temptation is to design the site, and access to the house, in a symmetrical, logical pattern that makes sense on paper, that looks as though it will work well. But you must be realistic about day-to-day operations like taking out the garbage, mowing the grass, shoveling snow off snaking walkways, and carrying groceries from the car to the kitchen via porch steps and decks where a handhold is required.

The design of Radburn, the first "car

town" version of the utopian, garden cities, is a classic case of social design and social practice conflicting in a residential housing plan. The first Radburn superblock, built in New Jersey in 1929, occupied a fifty-acre site with centrally located municipal facilities like schools and playing fields. A central artery fed culs-de-sac with the formal façades of row houses facing each other across the street. This spine and radiating rib design, motivated by a desire to separate people and cars, created mini-blocks split in half by the culs-de-sac. Each mini-block had a service road separating it from the next block. But this divider turned out to be a connector because the families used the back door, the kitchen at the rear of the house, and the backyard for activities that brought them in contact with their neighbors. Families facing each other, front entrance to front entrance across the cul-de-sac, had little contact.

Social orientation is important house to house, and also house to site. For instance, a recent *Housing Magazine* survey of new home buyers showed that almost 80 percent wanted their family room facing the privacy of a backyard, while fewer than 20 percent thought a similar orientation was important for the more formal living room. Nearly 80 percent wanted the kitchen to face the backyard, but only 40 percent thought the dining room should have that view.

These figures correlate to the home buyer's perceptions of the importance of activities within the house. At the top of the list are cooking and informal dining, the kind of family meals served up in an eat-in kitchen. Informal entertaining (the kind you do in a family room as opposed to a living room) ranked next on the list, supporting the importance of family room orientation on the site.

Social orientation is a complex part of a thorough site plan that will continue to work efficiently over your term of ownership.

The third orientation question—there are right and wrong answers to this one—is the physical interaction between the building and the environment. As a general rule, houses in northern latitudes should be placed to absorb the maximum amount of solar heat in winter, and to minimize the heat sink effects of wind exposure. Conversely, houses in southern climates should be oriented with major glass areas away from the sun, and toward the prevailing winds to increase cooling ventilation.

Everyone knows that wind can increase the effects of cold. This is reflected in the part of the winter weather forecast called the wind-chill factor. And when the summer sun makes you perspire, a breeze can cool you off by evaporating the moisture on your skin. But of these two factors that contribute to the thermal environment, wind is the easiest to control. Even in unprotected sites, trees or structural windbreaks can be used to ameliorate the effects of prevailing winds. Solar control, however, requires careful planning in the architecture of the house, and its orientation on the site.

If the sun could be turned on every day

like a light bulb, directly overhead, with the same energy-transmitting properties, solar orientation would be a simple matter. But in order to get the most free, clean heat in winter, and the least in summer, it is necessary to understand just a little bit about orbital mechanics, or the changing relationship of the earth and the sun.

The earth travels around the sun in an elliptical orbit every 365 and ¼ days. And it rotates on its axis about every 24 hours at an angle, or inclination. The orbit is roughly 3 percent eccentric, which produces a 7 percent variation in solar radiation during a six-month period. But this action does not cause seasonal changes. They are a response to the 23°27′ inclination. For instance, when the North Pole is tilted toward the sun, rays fall on the northern hemisphere at a steep angle that transfers heat efficiently.

In the northern hemisphere, at the time of the summer solstice, June 21, the days are long and this compounds the effect of summer heat because the sun shines longer each day. On June 21, the equator will experience a twelve-hour day. And as you go north to the Tropic of Cancer the days are slightly longer. Conversely, on December 21, the winter solstice, the North Pole is tilted away from the sun, rays hit the planet's surface at a low angle, the days are shorter, and, therefore, colder. So much for Astronomy 101.

The practical point of all this is that on a given day, at a given hour, a site orientation at one latitude must be different from the orientation at another latitude in order to make each one work efficiently. For instance, on February 1st at 2:00 P.M., in Miami (approximately 24° north latitude), the sun's rays will follow a bearing of 45° SSW and strike the house at an altitude angle of about 45°. At the same time in Seattle (approximately 48° north latitude), the rays will follow a bearing of 32° SSW at an altitude angle of 28°.

These differences should be taken into account to maximize winter heat from the sun in Seattle, and to minimize summer heat in Miami. For example, on a clear January day the south wall of the Miami house will be exposed to 75°F temperatures, and about 1,600 BTUs per square foot of wall area. But in July most of the solar radiation (about 1,200 BTUs per square foot) will come from the east and west.

Orienting a frequently used family room with a wall of sliding glass doors toward the south will make the room warm in winter and cool in summer. This orientation makes the room energy efficient by maintaining comfortable temperatures naturally, with a minimum of expensive heating and cooling in each season. But if the Miami house is in a development where, because of the bearing of an access road, the family room faces north, it will receive only 250 BTUs per square foot in January and require more heat support than other, less frequently used rooms in the house.

To analyze a site for solar orientation you should obtain an accurate survey, or compass readings taken on the spot. Bear in mind, though, that compass deviations from true north will affect orientation. To be accurate, consult an isogonic chart (available from the U. S. Department of Commerce and the U. S. Geological Survey if you can't get it in your local library), which shows west degree corrections for areas toward the East Coast, and the east degree corrections for areas in the Midwest and western United States.

This information can then be correlated with temperature statistics from the U. S. or local weather bureaus, and from design books like *Architectural Graphic Standards* (available in most libraries), that contain detailed tables on sun angles and solar radiation.

LEGALITIES

Obviously, there is more to site selection and building orientation than simply finding a nice-looking piece of land. But after you have found the right plot, and thought about how the building can be placed to look best, to work efficiently with your daily living patterns, and to use the minimum amount of heating and cooling energy, there is one more maze to work through. In most areas of the country many aspects of homebuilding, from the type and location of a septic system to the amount of land left uncovered around the house, are regulated by building, environmental, and health codes.

Before you build a house, before the plot is improved, before you spend time and money on any of the site orientation details, you must be sure that the basic positioning of the house does not conflict with local codes. A plot plan, submitted for approval before a building permit is issued, must accompany the detailed plans showing the construction of the house. On it, the surveyed boundaries of your property are noted along with prominent characteristics of the terrain, a stream or culvert for instance.

Be sure that the site you choose conforms to minimum front, back, and side yard allowances, and that it will continue to do so if you plan an addition for construction sometime in the future. You should also check for special easements or other infringements on your property such as a right-of-way for a utility company.

If you are buying in a development, code compliance is the responsibility of the builder. If you are running your own job, take a tentative plot plan into the local building department, and discuss code compliance with an inspector who is familiar with local geographic conditions, soil composition, the average depth of wells in the area, and other matters that determine exactly where you place the house on the site. Houses can be moved. But it makes more sense to devote the extra time and attention to site planning that will ensure a long-term, economically and socially efficient orientation.

3

MASONRY TOOLS AND EQUIPMENT

There is a big difference between working with masonry and working with wood. The characteristics of these two common residential building materials are very different, and so are the tools and techniques used to put them in place.

An average do-it-yourselfer can cut wood timbers accurately, quickly, and effortlessly with a sharp circular saw. But it is difficult to split two inches off an $8 \times 8 \times 16$ concrete block, and only a few construction companies have the resource of a water saw that can cut through concrete.

Masonry is heavy, abrasive, hard, and rigid. You can't glue bricks or cinder blocks together, file them down with a rasp, or plane one edge slightly to get a better fit. Masonry is also more permanent than wood and a lot less forgiving during construction. It is less adjustable to varying field conditions, harder to drill holes through, and more difficult to drive nails into than wood.

You can correct discrepancies in a finished wood frame by attaching cleats or braces to straighten out bowed joists and rafters, by driving a few sixteen penny nails into the bottom of a corner post to draw it into a proper, plumb position, and by adding thin shingle shims under a slightly miscut rafter. But once a poured concrete or block foundation is in place and cured, you have to live with it. If the corners of the foundation wall are out of square, you have to place the 2×6 sill slightly out of square with the masonry to compensate for the error.

Because masonry is more of an all-or-nothing proposition than wood, and a more monolithic material, the masonry tool box is a lot smaller than the carpentry tool box. Most of the finishing, trimming, and finessing tools are unnecessary—no saws, no planes, no rasps and files. But the absence of tools that correct mistakes and make fine adjustments does not mean that working with masonry—

Masonry Tools and Equipment

pouring concrete slabs and footings and building foundation walls out of concrete blocks—is an easy job. It may be less intricate than framing work, there may be fewer elements to deal with, and fewer connections between them, but if anything, masonry work is physically more demanding than framing work, where each piece of wood is lighter than each concrete block, and where the connections are made with nails instead of mortar.

This is one reason why many builders subcontract masonry work. While they may take a hand in building the wooden forms, and, if they are acting as general contractors, will certainly supervise the layout process, the efficient handling of masonry is frequently left to subcontractor specialists.

If you are planning a large project (a major room addition, for instance), it would be wise to ask for bids from at least three masons, not individual bricklayers but contractors who have equipment and a regular crew of workers. If you decide to do the framing work, or the roofing or Sheetrocking on your project, you can proceed at your leisure. But masonry work often cannot be strung out over several days. Once the forms are in place for footings, the concrete should be poured continuously, even if it takes several truck loads, to ensure structural integrity. And while modular concrete blocks can be placed one at a time, the mortar that bonds them into a structurally sound wall must be mixed in batches. Once you add water to the mix you are committed to using it. If you run out of gas the remaining mortar will be wasted.

As with many building processes, an essential key to success is good planning. And since the footing and foundation system of a house is crucial to the success and durability of the framing and finishing steps that follow, the planning for this part of any construction job carries extra importance. Not only will the masonry system carry the structural weight of the framework above it, but it will determine the starting points, the locations of the first masonry-to-wood transitions that are so crucial.

Once the footing and foundation system is in place the entire project changes from an idea expressed on a blueprint to a physical reality that occupies space. On paper, even on the most detailed plans checked and double-checked by highly qualified architects or engineers, structural and design elements are treated in two dimensions. And the translation to three dimensions inevitably uncovers gray areas and inconsistencies that must be solved in the field.

So the groundbreaking is a crucial building step but also an important emotional change from planning to execution. To ease this transition you should use the plans to prepare for the job in every way possible. Go over them again and again, noting the tools and materials you will require for each step. Make a rough time chart indicating the dates when materials must be ordered, and the tools and equipment you will need on hand to put them in place. Large construction companies keep track of the building process, and minimize delays, by drawing up elaborate flow charts that specify every building procedure, however insignificant.

Certainly it does not make sense to order concrete and then discover that you don't have the right kind of float to finish the surface. Assemble the tools and equipment you will need for the job before work begins, and since the initial excavation and layout for masonry footings and foundations influences every phase of the job, pay particular attention to the tools required for accurate measuring, plumbing, and leveling.

Here is a case in point that demonstrates how important this first stage of construction is. Most building materials are modular, i.e., they have consistent width, depth, and length dimensions that fit together into a structural system. If, for lack of attention or even something as simple as a slightly inaccurate level, the top of your foundation wall is out of level, the wooden sill attached to it will also be out of level. Of course the sill could be shimmed (made level by inserting different thicknesses of wood shingles or slate at various locations), but this would defeat the purpose of a sill, which is to provide solid, continuous bearing between the wooden superstructure and the masonry that supports it.

Once the initial error has been made, the efficiency of the modular building materials that are then applied will be lost. The 2 × 10 floor joists that rest on the sill will transfer the out-of-level foundation to an out-of-level plywood subfloor. Correcting the problem can add considerably to the time spent on the job: each floor joist can be notched where it rests on the sill, but this is an aggravating, hairsplitting process that can easily be avoided by attending to details during footing and foundation work.

Bear in mind that discrepancies in your initial layout will be transferred through each step of construction. If the floor is out of level, pre-hung doors won't fit properly in their frames, and panels of Sheetrock will have to be custom cut. But no matter how careful you are, a true foundation cannot be constructed with inaccurate tools. If you already have a 4-foot level, take it into a local hardware store or lumberyard and test it against new levels in the rack. Stack them on the counter and check the bubble readings in each vial. If your level is not reliable you should replace it. On wooden varieties the bubble vials are built-in, so the entire tool must be replaced. But on many aluminum levels the vials are screwed to the frame, making replacement possible.

The masonry tools and equipment you are likely to use for residential building projects fall into four general categories: tools for handling, placing, cutting, and, most importantly, for laying out the job accurately so that the work done will not require unsightly and structurally inferior repairs before the mortar is set and the concrete is cured.

LAYOUT TOOLS

Most tools that you will need to measure, plumb, and level masonry work can

Masonry Tools and Equipment

also be used on framing. There are only a few exceptions, and they have to do with the difference in scale between the fewer but larger masonry elements as opposed to the more numerous but smaller framing elements.

Starting from scratch on piers for a deck or a full perimeter foundation, the overall dimensions of the job must be established. You may get by with a 6-foot extension rule for a small job like a set of porch steps, but larger projects require larger capacity measuring tools to ensure consistent accuracy.

LONG TAPES If you measure foundation forms on a 48-foot-long wall with a 6-foot ruler, you have eight opportunities to make a mistake. Each time you start the ruler from scratch, mark its end, then start from scratch again, there is a possibility that the ruler will slip, that your mark may be a little off, that you will hold the ruler away from the mark in order to read it. Since these initial measurements are so crucial to the job you should go to great pains to keep the possibilities for error at a minimum. And in this case the answer is a steel tape, either 50 or 100 feet long, depending on the size of the job.

For best results, and extreme durability in all weather conditions, use a Mylar-coated tape with a permanently lubricated rewind bearing, sealed in a chrome-plated case. This is expensive ($15 to $25) but is built to last indefinitely. Rust is the chief enemy of measuring tools. It makes them difficult to read and can foul the rewind mechanism. A Mylar tape will not rust, even when laid over damp wood or masonry on a regular basis. Most have a ring attached to the end with an adjustable hook that facilitates reliable one-man measuring of house-sized distances.

Always pull the tape taut and flat before marking distances, and try to use the same tape throughout the job. Discrepancies may be small, but different tapes (particularly wood versus metal rules), will expand and contract at different rates according to changes in the weather.

SHORT TAPES For measuring shorter distances, the height of several courses of block, for instance, a 6-foot metal tape is convenient. An alternative for carpentry work, a wooden, folding extension rule, will not stand up as well to the rough treatment around abrasive masonry surfaces. Again, Mylar cladding will make the tape readable, and can be easily cleaned. Be advised: tapes advertised as "easy-to-read yellow" do not offer the durability of Mylar.

An important consideration is the width of the tape. For one-man measuring, the extra blade stiffness of a $3/4$-inch- or 1-inch-wide tape is worth the small increase in price. With a 1-inch tape you should be able to measure across an unsupported space nearly as long as the tape itself, a distinct advantage when you are working in a foundation trench, surrounded by tools and materials that impede free movement.

Short tapes also have a hook at the end, although it is shorter and less secure than the toothed hook common to long tapes. It is usually attached with one or two small rivets that eventually work loose. This $1/16$ or $1/8$ inch of play makes it important to pull the tape taut each time you make a measurement.

LEVELS Level footing and foundation systems are crucial. And to level them with the high degree of accuracy required the tools you use must be faultless. If a 4-foot level is off by only $1/8$ inch (that's only one of 384 $1/8$-inch divisions over the length of the level), the 80-foot-long wall you use it on may wind up a full $2 1/2$ inches higher at one end than the other. If the level is off by only $1/16$ inch the floor will be $1 3/4$ inches out of kilter.

You can see from these simple calculations that a short level increases the magnification of any error while a long level will minimize it. The most effective way to minimize the error would be to use one giant 80-foot level. If it were off $1/8$ inch the 80-foot wall would also be off only $1/8$ inch. Mason's levels, similar to carpenter's levels except for length, are available in 6- and $6 1/2$-foot lengths at most building supply outlets, and up to 8-foot lengths from commercial supply houses.

At these extreme lengths the price of an average carpenter's level, about $15 to $25, increases to $40 and up. But there are alternatives, some less expensive and only slightly less reliable, and one considerably more expensive and equally more reliable. First, you can increase the effective length of a standard 4-foot carpenter's level by gluing a 2 × 2 dimensionally stable hardwood strip to one side. A 6-foot-long board, proportioned to extend the level by one foot at each end, will be accurate if it is protected from the weather by several coats of spar varnish.

Water level This is the inexpensive alternative to a mason's level. It consists of a long, clear plastic tube (accurate at any length), that is filled with water treated with a red dye. Following the basic principle of physics that water always seeks its own level, this tool determines levels between two points. One end of the tube is held (or temporarily tacked) against one end of the foundation wall while the other is read against the forms or stakes or concrete block at the opposite end of the wall.

As long as all air bubbles have been cleared from the line, this tool is faultless. Its only drawback is that it measures point-to-point instead of continuously. In other words, it will not detect an uneven block wall that bellies up and down, only that the two ends are in or out of level with each other. This tool is very inexpensive and can be used year round as the dye contains a chemical antifreeze.

Transit This is the common term used to describe levels and surveying transits that incorporate magnifying optics. A contractor's dumpy level (cost is in the hundreds of dollars, making this a commercial tool) is leveled in place on a tripod. It can swivel 360° to cover any job site. It can be used with an engineer's rod (a long pole marked down to $1/100$ of a foot), or simply with a piece of wood, often called a story pole, with critical height dimensions of the job marked on one surface. Once the foundation forms are in place the story pole can be placed at any point along the perimeter and viewed through the eyepiece (most provide at least 18-power magnification). If the mark on the pole is above the eyepiece cross hairs, the form should be lowered. If the mark is below, the form must be raised. This tool combines the advantages of a mason's level and a water level. It is accurate enough to satisfy any residential limitation and can be used conveniently to check any portion of the construction. Although the cost is prohibitive for do-it-yourselfers, this instrument may be rented for the relatively short period of time when footings and foundations are under construction.

One inexpensive alternative to a mason's level is not acceptable. A line level, a small tube holding a single bubble vial, is hung on a string stretched between two points. Although the small tool is very lightweight, it can produce stresses in the string that can affect level readings. And the line level's short length (about 3 inches) offers the possibility of magnifying a tiny error into a catastrophe. Torpedo levels (about 9 inches long) are fine for leveling 12-inch shelves but not for house-sized distances.

SQUARES For masonry work where the accuracy of joints is determined either in the manufacturing process (for cinder and concrete block) or in the formwork (for poured concrete), a framing square, also called a rafter square, has limited value. This metal square, usually 2 inches wide with a 24-inch and a 16½-inch leg joined at 90°, can be used to check the accuracy of corners.

As interlocking concrete blocks are buttered with mortar and set in place, the square can be held horizontally against the inside and then the outside of the corner as a 90° guide. It should not be used to determine vertical alignment over several courses of block.

Again, following the principle of using a tool where its accuracy is most reliable, don't try to determine a 90° angle between two long walls with a 2-foot square. The first few blocks might be square, but this tool cannot measure alignment past this point.

When forms are constructed you can check for overall squareness using the principles of a 3-4-5 triangle, which are that a 90° angle will be formed if two legs of the triangle are in a proportion of 3 to 4, and the hypotenuse is 5. Using boards cut to length or dimensions carefully laid out on the foundation forms, you can check corner squareness by fitting a board corresponding to the hypotenuse leg of the triangle between the 3- and 4-foot dimensions on the forms.

A combination square with one 12-inch leg held in a cast metal handle is primarily a carpentry tool but can be used to mark splitting lines on concrete blocks.

LINES AND BLOCKS This tool provides an accurate and inexpensive way to check horizontal alignment, either on wooden forms or on block walls. The blocks are notched to grab the corners of masonry block, and grooved to provide a path for a string line. The principle is straightforward. If the corner blocks hold the string 1 inch away from the foundation wall, a block of the same thickness can be used at any point along the wall to check for the same margin. If, in the middle of the wall, a 1-inch block is placed against the wall, and the outside edge just touches the string that is stretched between corners, the middle of the wall is in alignment with the corners. If the block shows a ½-inch gap between its outside edge and the string, the wall is bellied in by that amount.

You can easily make your own set of blocks (hardwood, with a right-angle notch, sealed to prevent swelling), cutting several spacers of exactly the same depth from the same piece of wood. Used with a thin nylon (as opposed to cotton) line, this inexpensive setup can also be used to maintain a level line as you work between corners to fill in a course of block. However, each course should be leveled at the corners to eliminate discrepancies caused by varying thicknesses of mortar between courses.

PLUMB BOB Simply a weight on a piece of string, this tool can be used to align two points vertically. For most residential masonry work, however, vertical alignment can be accurately determined with a mason's level, or a 4-foot carpenter's level held against a truly straight length of 2 × 4.

HANDLING TOOLS

Because concrete is a dense and abrasive material, the tools and equipment used to mix it and transfer it from one place to another must be heavy duty. To preserve their effectiveness, always clean concrete tools thoroughly after each use and dry all metal surfaces to prevent corrosion.

CEMENT MIXER Mixing more than a few bags of concrete or mortar by hand is an arduous job. Even for a small project—like a retaining wall measuring 1 foot wide, 5 feet long, and 2 feet high, requiring 15 cubic feet of concrete (about half a yard)—you can save a lot of time and effort by ordering from a ready-mix company. Most will sell small loads, and with trough extensions off the back of the truck, they can place the concrete directly into the forms.

But for jobs that require a regular supply of small amounts of concrete or mortar, consider renting a motorized, gas-fueled mixer. Capacities range from 1½ to 3½ cubic feet.

WHEELBARROW A small, thin-gauge garden wheelbarrow will not provide the capacity or the strength for moving masonry. Although relatively expensive (about $65 to $80), a contractor's wheelbarrow is strong and stable enough to transport sufficient concrete block, concrete, and mortar to minimize interruptions in the construction process for more mixing.

A 5-cubic-foot watertight tank (usually 38 × 25 × 13 inches deep) can also be used as a mortar box. Transport of heavy materials is facilitated by a pneumatic tire, which cushions the load and maintains the thoroughness of the mixture of sand, cement, and water when it is moved. Hard rubber tires will transfer enough shock to settle the aggregate, creating a poor mix. This is why ready-mix trucks rotate the load in transit, and again when they are emptied.

SHOVEL AND HOE For hand-mixed batches of concrete and mortar, dry ingredients should be combined before water is added. This can be done with a shovel (a square-edged model of high carbon steel will get into the corners of a mortar box) or a mixing hoe, a heavy-duty garden hoe with several perforations to induce more thorough mixing. Long-handled shovels (about 48 inches long, ash) provide good leverage but require more hand strength to balance loads than D-grip shovels (usually 26 to 30 inches long with a closed grip perpendicular to the handle stem).

HAWK This tool is simply a flat piece of metal attached to a handle. On it, a modest amount of mortar can be carried from one block to another as you work. It is easy to make your own from a 12 × 12-inch square of ½-inch plywood, screwed to a piece of 2 × 4 tapered to fit comfortably into your hand. If you opt for the manufactured version, a standard hawk is 13 × 13 inches and should be made of at least 12-gauge, hard-rolled aluminum. For proper balance, a 5-inch stub handle should be attached to the bottom of the hawk with a reinforced circular flange.

PLACING TOOLS

A small, inexpensive collection of trowels and floats will suffice for almost every masonry job. Many styles of each tool are available. For instance, a standard mason's trowel (also called a brick trowel) is tapered to a point at the end, reaching 4 to 5 inches across at its widest section, and extending from 10 to 13 inches in length. A Philadelphia pattern is gently tapered at its base, while a London pattern is cut more in the shape of a teardrop. These distinctions are a matter of personal preference and do not affect work quality.

TROWELS A mason's trowel (12 or 13 inches × 5 inches) and a pointing trowel (about 6 or 7 × 3 inches and more pointed than a mason's trowel) will serve to butter concrete blocks with mortar, to shave off the excess pushed from horizontal joints by the weight of the block, and to smooth both flush mortar joints and the tops of blocks where the cores have been filled. They should be forged in one piece from high-grade trowel steel, tempered for extra strength, ground, and polished.

Finishing trowels are rectangular (11 × 4 inches up to 16 × 5 inches). They can be used in floating concrete, also called puddling, which brings a watery cement mixture to the surface, and to smooth the surface. High-quality finishing trowels are made from high-grade spring steel with either an aluminum alloy or malleable steel strip mounting secured to the blade with rivets. A gently curved handle, sometimes referred to as a California handle, may provide more controlled articulation of the blade.

FLOATS Designed specifically for

puddling, floats may be made of aluminum, wood, or either a cork or sponge surface to produce a variety of finishes. Aluminum will give a smooth, icelike finish, while a sponge-surfaced float can be used to create a slip-resistant surface on concrete steps. Most float handles are U-shaped (attached to the blade at two points) as opposed to trowel handles (L-shaped; attached at only one point).

DARBYS These tools, made of wood or aluminum, are floats that cover a large surface area with one sweep. Models are available in 4-, 6-, and 8-foot lengths. Shorter models have a continuous raised wooden strip for a handle, while longer versions have an extension arm so the float can be manipulated from an upright position.

Once the concrete surface has been puddled and set up, a hand float or finishing trowel can be used to achieve the same purpose but not the same results. On large, open areas, your feet will leave imprints in the concrete as you use the hand float. You can minimize the effect of your body weight (it will affect the density of the mix) by working from kneeboards, 2-foot-square sheets of plywood that distribute your body weight more evenly over the surface of the fresh concrete.

Darbys, however, will produce more uniform results over large, open areas because the size of the blade matches the scale of the surface. But the most reliable float and leveler is a straight length of 2 × 4 cut to ride on edge forms, called screeds, that border the concrete surface. If the screeds are level, and the 2 × 4 is straight, you can't go wrong.

EDGERS These are small tools, about 6 inches long, with two-point handles attached to a molded blade. Shapes are available to smooth inside and outside corners on steps and to create either hard or soft edges.

JOINTERS These are small, rectangular tools with a two-point handle used to trace grooves in newly poured concrete surfaces. The outside edges of a jointer are turned up slightly to avoid gouging the finished surface. Down the center of the blade a V-shaped tongue protrudes to cut the joint. This tool should be guided against a straight board laid flat on the concrete. The resulting groove, called a control joint, is used to relieve tension stresses (or at least to minimize their damaging effects) in the surface of concrete. If the slab heaves or settles unevenly, and most do, tension cracking may be confined to the control joints where its effect causes less deterioration and its appearance is less obvious.

RAKERS Aluminum raking tools are cast in different shapes to create patterns in mortar between courses of concrete block. The thin, narrow tools are angled so that your hand will not scrape against the masonry surface as the tool is drawn along the joint. V-shapes, convex rounds, and flats are common shapes, but you can use any piece of plywood or heavy gauge metal cut to these or other patterns.

CUTTING TOOLS

Short of using a water saw (an expensive, water-lubricated saw that will make agonizingly slow but clean cuts through masonry materials), most cutting work in concrete block is done with hammer and chisel. Obviously, conventional wood chisels, made to hold a razor-sharp edge, would shatter under this kind of use.

The principle of cutting masonry is to score the surface with hammer blows on a brickset or stone chisel, a process that weakens the block in a uniform, controlled manner so that a final blow with the hammer breaks the block in a relatively straight line. This same principle is used in cutting glass.

BRICKSET AND STONE CHISEL Bricksets have the stem of a conventional cold chisel with a widened cutting blade ranging from 3 to 4 inches. Forged tool steel is required to maintain a reasonably sharp edge, and to resist chipping and splintering. For smaller cuts, or where a brickset cannot deliver enough cutting power, a more compact stone chisel, with a blade about 2 inches across, can be used.

HAMMERS Most masons carry a bricklayer's hammer, a square-faced hammer that tapers to a chisel end on a straight claw and weighs 18 to 24 ounces. It is an effective, all-purpose tool for masonry work that can be used to deliver blows against the stem of a brickset or to score block directly with the chisel end.

A drop-forged, steel stonemason's hammer (weighing up to 2 or 3 pounds) will certainly deliver more force but it will also wear out your arm in a hurry.

DRILLS Holes can be drilled with specially made masonry bits. They resemble twist drills although the spiral edges are not sharpened. The cutting is done at the tip of the drill, preferably by a carbide, V-shaped insert. A powerful $3/8$- or $1/2$-hp drill is needed for most holes up to $1/2$ inch in diameter. For larger holes a star drill must be used. It has the shank of a thin cold chisel, flared at the cutting end into a star pattern. As each hammer blow is made the drill is rotated by hand to set the cutting flanges at a different angle. Since this is a time-consuming process, hardware should, if possible, be limited to attaching lags and bolts of $1/2$-inch diameter or less.

MISCELLANEOUS HAND TOOLS In addition to a standard assortment of general-purpose tools like pliers and screwdrivers, you may need a hacksaw and a heavy duty wire cutter. A hacksaw is used to cut reinforcing rods and to cut sill anchor bolts that protrude too far above the sill surface.

Wire cutters are required to trim reinforcing mesh that is laid just above the base of concrete slabs. A bolt cutter with double-acting jaws, capable of cutting medium hard steel up to Rockwell C31 hardness, will easily cut 6×6-inch welded-wire reinforcing mesh.

Since masonry is a hard, abrasive material it is advisable to wear protective glasses whenever you use a hammer and cold chisel against a finished surface. This same precaution should be taken when driving fluted masonry or cut nails into concrete or block.

FOOTING CONSTRUCTION AND CONCRETE FUNDAMENTALS

Several years ago a wire service photograph made a lot of papers across the country, not because of its news value, but because it made everyone who saw it do a double take. The bird's-eye view of a construction site looked commonplace at first glance. The caption described the open excavation site as the future cellar of a large home in a Chicago suburb. The picture showed wooden forms filled with concrete bordering a rectangular dirt cellar area, clear of debris and ready for a concrete floor, except for one obstruction—a large, diesel-powered backhoe.

Somehow, it was left there as ramps were removed, as the foundation wall forms were erected, as the forms were filled with concrete, as bracing was removed, as the crew ate their lunch. The caption did not specify how the backhoe was removed. The builder may have been too embarrassed to talk about it, because this is the excavation equivalent of painting yourself into a corner.

Small errors, just an eighth of an inch here and there, can add up to make a structural system imprecise and ragged around the edges. A single large error, a planning or design blunder on a scale of the backhoe in the cellar, can make it impossible to continue. Very few steps in construction are superfluous. Almost all depend in some way on preceding steps: subflooring on the joists, joists on the sill, sills on the foundation wall, right down to the ground.

The importance of these structural dependencies is particularly crucial in footings. They support the house. If they have a built-in weak link, all the care and precision applied to the framing and finishing work they carry may be wasted.

For most residential applications footing construction is a five-step process: excavating, preparing the soil the footings will rest on, laying out the building lines, constructing the forms, and pouring the concrete. For a simple footing you can

Footing Construction and Concrete Fundamentals

dig a trench, pour gravel along the bottom, and fill it with concrete. If the trench is about a foot wide and two to three feet deep, and if the concrete is properly proportioned and mixed, and if the trench area is, and will remain, relatively dry, chances are that this oversimplified procedure will provide all the strength you need—particularly if you take the trouble to place reinforcing rods a few inches above the floor of the trench. So why bother with anything more complicated? To eliminate the deterioration, the maintenance, and the structural catastrophes that can occur with the aforementioned procedures.

Essentially, a footing is a large, heavy, rectangular mass of concrete connected strongly enough to be an immovable object. If you placed a massive structure like this on top of the ground it would sink through soft topsoil, crack, probably break, and tear apart the building it was intended to support. So the first step is to dig down to solid subsoil, below the level where the heaving action of frost can do any damage.

EXCAVATION

Before any earth can be moved you must translate the building location from the plot plan drawings to the actual location on the site. Corners of the house can be reliably established when the property is surveyed, and from measurements off locations noted on the plan as initial points (IPs) or monuments. These are permanent surveying markers left on the landscape, such as a concrete pier with a metal disk on top showing precise bearings. If a survey has been made recently, fluorescent orange flags, used for alignment and marking by the survey team, may be visible on trees about the property borders, to help you fix compass bearings.

However, if you have any doubts about the true location of the building lines, get professional help. On new houses and additions to existing homes, building lines a foot off one way or the other may not change structural integrity as long as they are all off the same amount in the same direction. But transposing the house even a marginal distance can get you in trouble with the building department.

In Houston, where air rights, or the air space directly above a building, are very high priced, a commercial builder was required to remove less than six inches of concrete ledge from an upper floor that intruded into the expensive air corridor of an adjacent building. Your local building department can require you to dismantle (literally tear down) part of a structure if it intrudes on a legal right of way, and even if it violates proportional yard allowances.

Of course, the chances of obtaining a variance for a six-inch imbalance of side yard proportions are excellent, but you should still take the time and spend the small amount of money required for making a variance appeal in advance. Generally, this can be done by filing a copy of your building plans with the appeals board, and posting certificates adjacent to your property announcing your application for a variance. Theoretically, this gives your neighbors a chance to lodge a protest. And remember when you are looking at property that, in most cases, building design and layout can be made to conform to physical limitations. But in some cases, a particularly long, narrow plot for instance, the design you have in mind may only be possible to build if a variance is obtained.

The next step is to lay out the lines of the house foundation wall, but bear in mind that excavation work, particularly for a full cellar, is done most efficiently by machine (usually a backhoe), which must have access to the site. Complete site preparation—that is, digging, removing earth, and grading—may require a backhoe, dump truck, and a small bulldozer. And in many cases the operators of these machines are not terribly careful about the path they cut in and out of the site. Plan this route carefully, marking any trees that you want to preserve with bright tape. Occasionally, trees close to the site that are most likely to be scarred are protected with old tires tied around the trunks. In all but extremely unusual plans it makes sense to clear and rough grade the driveway first. This area can then serve as access to the house site and as a preliminary storage area for bulk material deliveries.

Layout

There is an obvious conflict between marking lines of the house so they can be accurately read and having the freedom to dig efficiently without disrupting the careful layout. To accommodate both concerns a system of batter boards is used. At each corner three stakes made from 2 × 4's are cut to a point at one end with a circular saw and driven deeply into the ground to form the three corners of a right triangle. The stakes are set at a minimum of four feet outside the line of the foundation, and joined together horizontally along the legs on either side of the right angle with 1 × 4's (Figure 6).

Each batter board corner encompasses the point where the foundation lines meet so that continuations of the layout lines (four feet away from the actual digging site) will cross the 1 × 4 horizontal boards attached to the stakes. To establish square corners for the foundation wall, first drive a small stake at the corner of the building, and use a long steel tape for triangulation. At each corner, using the relationships of a 3-4-5 triangle, you can lay out a 9-foot and a 12-foot leg (for instance), checking triangulation with a 15-foot hypotenuse, although any 3-4-5 proportion will work. For extreme accuracy—a thing you can't have too much of at this stage of construction—use 2 × 2 stakes neatly cut with a clean surface area. Then, when triangulation measurements have been taken and checked, and rechecked, mark the exact points of intersection between the triangle legs with small nails.

Before you start digging, and even if you would bet your mortgage that the corners of the house are square, make one final check. Carefully measure the diagonals of the house (see Figure 7—if the

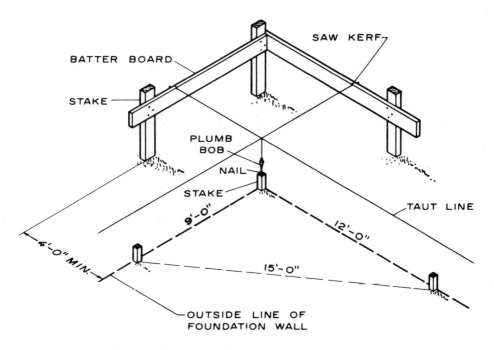

Figure 6. Batter board layout for perimeter excavation

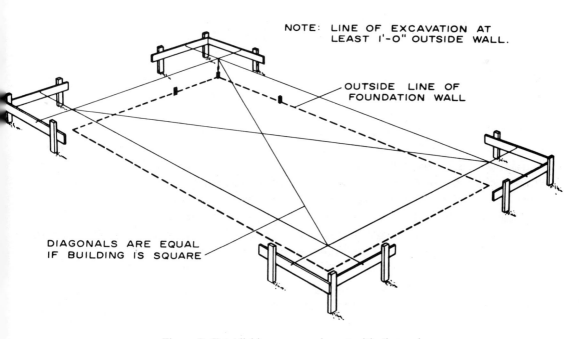

Figure 7. Establishing a square layout with diagonals

shape of the foundation is irregular you can square off a few corners with temporary 2 × 4's to create a rectangle). If the building is square the diagonals will be equal. If, for instance, the diagonals of a 30 × 40-foot foundation measure 50 feet, 1 inch and 49 feet, 11 inches, respectively, you should go back to the corner layouts to find the error or consider triangulating in 30-, 40-, and 50-foot dimensions.

When the corner stakes on the foundation line have been properly squared, find the continuations of the layout lines on the batter board horizontals by stretching a line against the outside edges of the nails driven into the 2 × 2 stakes. At this point mark the outside lines of the foundation wall on the batter board in pencil, then cut saw kerfs along the top edges of the boards to hold the layout line in place (Figure 6). Remember that a slight error can be magnified into a serious structural fault when compounded over the length of a wall. After the lines have been placed, recheck the corner triangulation and measure the diagonals.

Batter boards can also be used to establish and control the finished height of the foundation wall. Granted, this is somewhat trickier because the 2 × 4 stakes must be driven into the ground bit by bit as level lines are established corner to corner. And it is not easy to adjust batter boards for exact depth while maintaining precise squareness. Nevertheless, this double or two-dimensional layout (for both horizontal and vertical control) is the best way to proceed.

Water levels can be used corner to corner and to check diagonal level. Most water level kits include spring clips for attaching the hose line to batter boards just for this purpose. It should also be clear that the determination of square and level points (two at each corner) is extremely valuable information that would be difficult to duplicate if the batter boards were disturbed. Placing them four feet away from the foundation line helps to keep them out of harm's way, but they must be solidly constructed in any case. Use knot-free 2 × 4's, pointed severely (at 60° or more) so they can be driven into the earth with little resistance. For extra strength the 2 × 4's can be backed up with 1 × 4 stakes tacked at one end to the side of the 2 × 4's, and driven into the ground to form the hypotenuse of a triangle with the 2 × 4 stake as one leg and the ground as the other.

As measurements are taken for triangulation and level readings are made for depth control, the horizontal batter boards should be adjusted with clamps. For instance, if a level reading shows that one end of the board should be raised ½ inch, back the clamp off slightly (enough to provide enough holding power to keep the board in place, but not so much that the board cannot be moved with a firm tap of a hammer), adjust the board, and then tighten the clamp again. When the series of fine adjustments have been made at each corner of the foundation, and only after depth and squareness have been checked thoroughly, the boards may be nailed in place.

Even if the vertical 2 × 4 stakes have

been backed up with ground braces, the process of nailing on batter boards can disrupt the layout. To overcome this, even if you are working by yourself, gently set three wire nails in a staggered pattern into the face of the 1×4 or 1×6 batter board while it is held in place with clamps, then hold the head of a sledge hammer against the back of the 2×4 stake directly in line with the nails. This extra weight will absorb the blows of the hammer and leave your careful layout intact.

At this point the obvious question becomes: Is the initial foundation layout worth all this nitpicking over fractions of inches? A case could be made for speeding up the construction process by making a relatively accurate layout, getting the concrete in the ground, and then exercising more care as foundation wall forms are built (if the foundation is poured), or as block is laid (if the foundation is unit-built). Sure, if the footing is 12 inches wide, it should be able to accommodate an inch or two of misalignment. Of course, the weight of the foundation wall will not bear evenly on the footing, which may lead to premature cracking, water infiltration, deterioration, maintenance time and money, and so forth.

So the next question is: Why uncover such an unpleasant can of worms just to save a little time on layout that you will most likely spend on squaring and leveling the foundation wall anyway? Once four level, square corners are reliably established to a fine degree of accuracy, measurement after measurement can, in turn, be reliably taken from these points.

Design Alternatives

At this point, before excavating, you will want to choose one of three possible foundation designs: full-height walls for a full cellar, short walls for a crawl space, or a slab on grade, which provides no cellar or access below the first floor.

Full cellars with 8-foot-high poured or block walls economically double the first floor area. You have to dig trenches for footings in any case, and by increasing the scope of excavation, and enclosing the area with foundation walls, valuable, livable space is gained without covering an excessive amount of the site. The chief drawbacks, perennial dampness or, worse yet, outright leaking, can be prevented with a thorough combination of drain tile, insulation, vapor barriers, and waterproofing applications before backfilling (discussed in Chapter 5).

Short walls, either poured or unit-built, can be used to keep the lowest first floor timbers at least 18 inches off the ground, creating a crawl space (literally just enough space to crawl into if you need to replace something like a broken pipe). This design minimizes excavation work, cuts down on block construction, and dispenses with most of the problems associated with a below-grade cellar, although the crawl space must be dampproofed and insulated.

Another alternative, slab-on-grade construction (see Chapter 6 for details), is a quick, economical way to prepare the foundation and the first floor, but you may pay for the construction conveniences with increased maintenance later on. With slab construction there is no space for heating pipes or ducts (radiant heating built into the concrete is outmoded), no access to the foundation wall should structural problems develop, and no dead air space between the ground and the house.

Thickened-edge slabs combine the foundation wall and the first floor in a single concrete pour, tied together with wire mesh reinforcing. A conventional slab on grade rests on the ground as well, but is carried around the perimeter by an independent foundation wall that can be made of poured concrete or block. Generally, sloping sites and low areas that puddle easily during rain are not suitable for slab construction because excessive time and money must be spent on drainage and waterproofing. Thickened-edge slabs, perhaps the most economical method, is particularly suitable in warm climates with shallow frost penetration.

There are many choices, using different materials and designs, including plans that combine various methods of construction to tailor the foundation to the site. For instance, on a sloping site the lower section might be fully excavated to provide garage and utility space, while the higher section of the site might be excavated only for a crawl space, providing enough room to run pipes and ducts from the cellar furnace to the rest of the house.

Cut and Fill

An economical proportion between cut and fill is essential to efficient road building. As the highway passes a hill, earth is removed to decrease the gradient, then used to fill up the valley in front of the next rise. If the route is planned properly the road builder will not have tons of earth to dispose of and will not need to buy tons of earth for fill. This ratio is less critical on a house-sized building site, but it should not be overlooked. Clean fill is expensive. So is the cost of trucking it in, placing it, and grading it.

Excavating a full cellar measuring 60 × 40 feet with 8-foot walls (assuming 1 foot above grade), means removing 16,800 cubic feet of earth, nearly 625 cubic yards, or enough to spread 4 inches of new fill over a football field. Applying the principle of proportioning cut and fill, it might be wiser to plan a final grade sloping from a height of 7 feet on one side of the foundation, down to a foot or two on the other side.

Whatever the lay of the land, it is good practice to use the highest elevation around the perimeter of the foundation as a control point, or high limit of the wall (Figure 8). For proper drainage, whether you excavate a full cellar or a minimal crawl space, finished foundation height, i.e., the lowest elevation of wood framing, should be at least 8 inches above the finished grade (Figure 9). And on many sites where there is no natural rise on which to build, the original grade may not slope away from the foundation on all sides, which is important for draining water away from the house. In this case, if only 8 inches of masonry wall has been left exposed above the original grade, fresh fill needed to create a slope will run very close to the foundation sill, a condition that encourages deterioration from wet rot and termites, even if a termite shield is installed. So keep this in mind when marking the point of highest elevation.

For minimal crawl space excavation, the finished grade should be at least 18 inches below the bottom edges of wooden floor joists, or, if a girder carries a split floor joist run, below its lowest edge.

Digging

Excavation can be accomplished with a front-end loader or a backhoe (also called a power shovel). In both cases the articulation built into the digging arms will create a soft slope from the bottom of the cellar floor to the finished grade. This area, called the back slope (the final grading against the finished foundation wall is called backfilling), should permit enough room to stand in the excavation outside the footing perimeter.

The precise angle of the back slope must be determined during excavation, or beforehand, if test bores have revealed subsoil type. If the soil is stable, like clay, the back slope may be close to vertical. If the soil is sandy, an incline approaching 60° or less may be necessary to prevent caving. If the excavation work is done by a contractor (either a subcontractor you hire and supervise or one hired by your general building contractor), you may find him reluctant to prepare an exacting batter board layout. But even if the contractor intends to roughstake the site, excavate, and lay out the building wall on the footing surface, the floor of the pre-

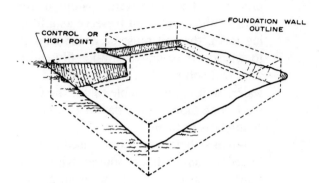

Figure 8. Equalizing cut-and-fill requirements for efficient excavation

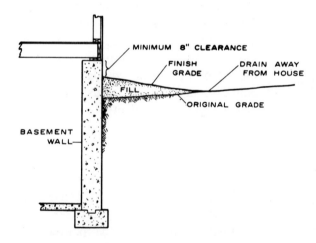

Figure 9. Establishing a finished grade for adequate drainage

liminary excavation should be carefully prepared. If a sharp trench is cut in the floor of the excavation to serve as a footing form, the soil must be dense enough to withstand caving. Additionally, most soils will soften when exposed to air and water, so it is not sensible to cut the footing trench more than a day or two before concrete is ordered.

When slab construction is used (a reinforced concrete floor poured with continuous bearing against the ground), perimeter footings and isolated, pile-type footings for lally columns (page 57) are still required. And in most cases the basic principles of excavation continue to apply. For instance, the finished slab surface should be at least 8 inches above grade, and the footing, whether reinforced and poured simultaneously with the slab (a thickened-edge slab), or poured independently, must still reach below average frost depth.

SOIL PREPARATION

Footings transfer the entire weight of the house to the ground, and consequently must rest on solid, undisturbed soil. Footings should never be poured over loose fill, which will compact unevenly; and if trenching is too deep or rock removal leaves irregular cavities in the bottom of the trench, the excess space should be filled with concrete and not leveled with soil.

If solid rock is encountered, it should be prepared by roughening and thorough cleaning. Loose rock, scaly coatings, organic deposits, and other foreign matter must be removed. Cleaning may be accomplished with steel brooms or, with commercial grade contractors, by water jet or wet sandblasting.

If earth is encountered it must be excavated so that the bottom of the footing is below the frost line. The map (Figure 10) shows the average depth of frost penetration across the country; however, local building departments should also be consulted to determine conditions in your area. Frost action is a result of freezing water. Sites with poor drainage may be particularly susceptible to heaving action as natural groundwater freezes. Some areas of northern Maine and Minnesota require footings up to 6 feet below grade, while most of Florida and southern Texas require only a foot of depth to avoid possible frost action. In every case, though, consult local building codes, which generally mandate that the bottom of the footing be at least 16 inches below grade.

If a soft subsoil like silty clay is uncovered, 4 to 6 inches of consolidated, granular fill may be added under the footing to reduce horizontal movement of the wall. Positive resistance to slipping may also be accomplished by adding a key (also called a lug), which protrudes below the center of the footing (see Figure 13, page 51).

Compressible soil is best improved by impressing a layer of sand or gravel into its surface. This follows the operating principle of improving soil-bearing capacity, which is to decrease void volume. That is, when particles that make up the soil have space around them, they have

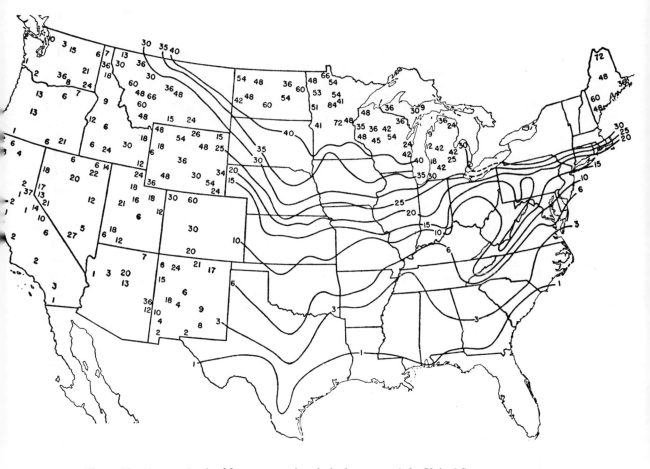

Figure 10. Average depth of frost penetration, in inches, around the United States

room to move into when loads are applied. But when these voids are closed, or at least decreased, the particles have no where to go; they support each other more directly, and consequently provide better bearing.

This process is centuries old and grew out of the need to locate strategic structures on strategic sites, and not necessarily sites with adequate subsoils. Soil consolidation was first accomplished by driving 6- to 10-foot wood piles by hand into soft ground. As soil consolidation technology improved, however, piles were driven mechanically, and the spaces filled with concrete, initially not to carry specific or concentrated loads but to improve the overall bearing capacity of the site.

Simply digging deeper, or increasing footing depth does not substantially improve bearing capacity unless a more stable soil is encountered. Roughly 80 percent of settlement that can cause structural faults and excessive cosmetic maintenance throughout a house results from compression of soil within a depth of only 1½ times the footing width. Unwanted soil consolidation may occur as well from adjacent loads, although this is unlikely in isolated, single-family house construction. But as a safeguard the footings for independent columns should be a minimum of 2 feet apart.

Failures of foundation walls caused by excessive settlement, heaving, or in some cases, lateral displacement (slippage) are rarely the result of footing failure. More often, the structurally damaging movement can be traced to incorrect design assumptions about the type of footing subsoil, or its uniformity. If one end of a house is carried on ledge rock with well-secured step footings, and the other end is loaded on unconsolidated silt, the foundation wall is likely to separate dramatically, and in a short time after construction, where the ledge ends and the silt begins.

Remember that even compact, hard, and dry clays will not remain in that state if excessive groundwater is present. Similarly, if foundation walls are backfilled with loose, moist fill that shrinks drastically as soil water is suctioned into the adjacent concrete, finished grades around the foundation can be destroyed, and voids between the dehydrated soil and the wall may fill with rainwater.

For solid construction requiring only minimal maintenance from some inevitable amount of settling, footings should be poured on solid, undisturbed soil. In special situations, for instance, where collapsible forms that permit heaving may be used over expansive or fat clays, protect your investment in the building by consulting the local building department about soil types, foundation designs that have proved suitable in the area, and if necessary a licensed foundation engineer.

FORM SIZE AND DESIGN

In normal soils, footing size is related directly to foundation wall thickness as follows. Footing depth should be at least equal to wall thickness, and footing width should be at least twice the foundation wall thickness, i.e., it should project equally beyond each side of the wall by one half the wall thickness. (Remember that the depth of the footing itself is completely different from its depth below grade, which is determined by the average frost penetration.) Of course, this general proportion of footing size may be superseded by local codes, and, when unstable soils are encountered, by specific computations of soil bearing value and the combined dead load of the structure.

While final loads are static (the dead weight of construction materials), footing forms must be able to withstand both the dead load of concrete, and the live load generated by the operation of pouring fluid concrete. On standard footings (8 inches deep by 16 inches wide poured on level, undisturbed soil), construction grade 2×8's can be laid on the excavation

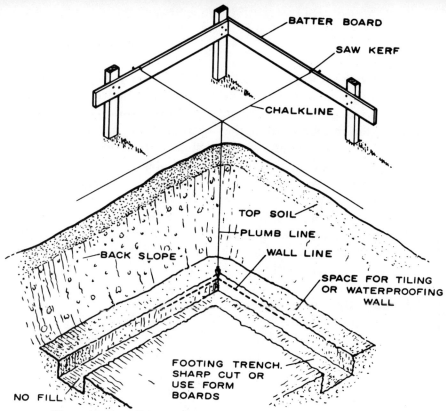

Figure 11. Establishing corners for footing trenches from layout lines

floor to hold the concrete in place. These form boards can be held in line with 2 × 4 stakes driven well into the ground with the long side (3½ inches actual) at right angles to the 2 × 8. In most cases, where a sharp cut trench is not feasible, 2 × 2 stakes, which are more practical for nailing, may be used at 2 to 3 feet intervals, providing they are backed up by compacted fill.

Corner stakes can be located by dropping a plumb line from the intersection of batter board layout lines above grade (Figure 11). But remember that the stakes will be driven so that their inside edges touch the outside edge of the 2 × 8 form. Consequently, the stakes should be

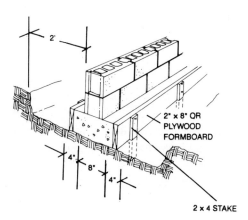

Figure 12. Typical form for footings

placed 1½ inches outside the final width of the footing. Also bear in mind that a point plumb with the batter board corner lines is actually the outside edge of the foundation wall, not the footing, which will project beyond the wall by a distance equal to at least half the wall's thickness.

After the critical corner stakes have been located and driven, brace them with 1 × 2's angled into the ground, and then connect the corners with a taut line. Intermediate stakes can then be driven at 2 to 3 feet intervals (depending on the depth and wall angle of the footing trench), by aligning each one with the line. Again, remember to account for the difference between building line and footing line. This is not an uncommon oversight. Somehow, as with the backhoe marooned in the cellar excavation, it is easy to get bogged down with small procedural details and overlook the obvious.

When all stakes are in place (and braced if trench wall support is minimal), establish the footing elevation, again, not from the floor of the trench, which will inevitably wander up and down hill a bit, but from the control point on the batter boards. As construction proceeds it becomes increasingly obvious how valuable it is to have reliably accurate horizontal and vertical layouts above grade. The form boards should be nailed to this mark and secured with double-headed nails. (They have the holding power of fully driven common nails but an extra head to facilitate form stripping.)

If form boards seesaw over trench floor high spots, cut the soil down with a sharp spade and remove it. Don't spread it over low spots in the trench floor. And if the footing area has many depressions, include a rough guess of the additional cubic footage in your concrete order. (For more about ordering concrete, see page 55.) But don't rush to the phone to place the order. Take the time necessary to check the consistency of form board elevations. Put yourself in the place of a demanding inspector. Try to find problems, search out inconsistencies. Now, before the concrete is poured, you have a chance to correct them.

Finally, the forms must be strengthened against the pouring force of concrete, particularly if the site is accessible only from one side, and extensive troughs are used to feed concrete into the forms. A case in point: On a site at the bottom of a steep hill, too steep for a ready-mix truck, foundation forms for a monolithic pour including footing and foundation wall were built and braced. Bad weather caused a postponement of the delivery for several days, and during this time a substantial amount of floor and wall framing was completed. Troughs were built from the road down to the site, and on the appointed day the ready-mix truck deposited about four wheelbarrows' worth of concrete in the chute. Nothing happened. The same amount was released again to push the first batch down the hill. Nothing happened. Not yet discouraged, the driver then released the balance of the load. The troughs creaked. The leading

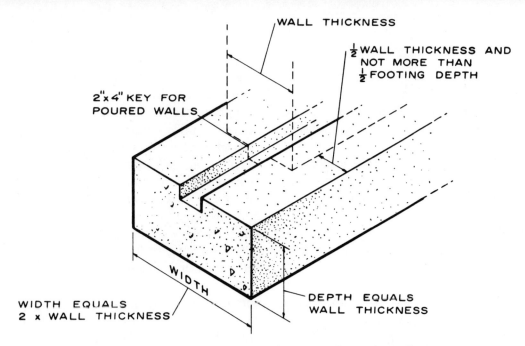

Figure 13. Keyed joint between footing and foundation wall

edge of concrete jolted a few feet down the steep trough, halted, and then shot down the hill toward the building site at an alarming pace.

Extreme conditions cannot always be foreseen, but the sea of concrete, splintered form boards, and collapsed framing that resulted in this case could have been avoided by better bracing. For a more conventional pour, nail spreaders made of 1 × 2's across the footing form at equal distances between the 2 × 2 vertical stakes. Connecting the top edges of the form boards in this way prevents splaying as the concrete is run in.

Keyed Footings

A common source of basement and crawl space water leaks is the joint between footing and foundation. This joint is eliminated in continuous footing/foundation wall pours (see page 62), and can at least be minimized by forming a key slot into the footing surface (Figure 13). This can be accomplished by centering a 2 × 3 strip along the length of the footing trench, flush with the top of the form. This location is arrived at by nailing the 2 × 3 to the spreaders, clinching the nails, and then turning the assembly over (with the 2 × 3 down), when the spreaders are nailed to the form boards. This assembly will be removed, along with the forms if possible, after concrete has been poured and set up.

Reinforcement

Steel reinforcing bars, placed about 2 inches from each side of the form and about 2 inches off the footing trench, will substantially increase the flexural strength of a concrete footing. Three bars (either $3/8$- or $1/2$-inch diameter) should be used in standard, 16-inch wide footings (the third bar is run down the center of the form), supported on split brick, and secured with tie wire. Rods come in 20-foot lengths and can be joined with three series of wire ties over at least an 18-inch overlap.

When rods are supported on block or brick sections, care must be taken not to disturb their location during the pour. A reinforcing rod pushed to within $1/2$ inch of the footing wall is relatively useless. Stability can be increased in two ways: first, by using commercial rod holders, suspended at intervals from the top of the forms; and second, by drilling fine holes through the sides of the form boards (at the 2-inch elevation off the trench floor), through which tie wire may be secured, and wrapped around the brick-supported bar.

At the footing corners, reinforcing bars may be bent (not crimped or grooved to form a precise 90° angle), so that each 20-foot length turns the corner a minimum of 18 inches. This provides a 36-inch overlap and increases the monolithic stability of the footing system.

POURING CONCRETE FOOTINGS

On small jobs concrete may be mixed at the site, either in a mortar box or in a rented, gas-fueled mixer. For large projects, and certainly for footings on additions or new construction, concrete should be ordered ready-mixed, i.e., proportioned and mixed at the plant and brought to the site by truck. This service is available in most areas of the country, even in remote, rural sites, although it is likely that additional transportation charges will be incurred for long hauls. However, a ready-mix order provides uniform concrete, and, if you order carefully, assures a continuous pour. Pouring only one side of a footing or foundation wall at a time is not acceptable.

To digress, concrete is a fluid mixture composed of sand, gravel (called aggregate), Portland cement, and water. Obviously, by combining these ingredients in varying proportions, a similar variety of characteristics will be produced in the finished concrete. Too much water produces scaling, a crumbling of the surface layer. This results from freezing and thawing before proper curing as well. Too much cement, which produces an over-rich mixture, causes crazing, a weblike tangle of hairline cracks resulting from tension cracking during abnormal shrinkage. Aggregate containing organic matter or lumps of clay can produce popouts (small craters in the concrete surface) as these elements are exposed to water and washed away.

By carefully altering the proportions, you can obtain specific compressive strengths to meet precisely calculated load demands. But for large or small projects requiring concrete footings, certain minimal standards should be maintained. Frankly, there is no way, short of elaborate field testing, to determine if you are receiving a specified proportion from the ready-mix company. However, these plants generally adhere to the recommendations of the American Concrete Institute (ACI) and have little to gain by substantially weakening the mix.

Concrete Mix Ratios

Concrete is generally measured and sold by the yard, really the cubic yard, or enough concrete to fill a 3 × 3 × 3-foot cube. A cubic yard holds 27 cubic feet, although ready-mix orders are generally subdivided into halves, and in some cases, thirds of yards. Ready-mix orders are conventionally ordered by the number of bags of Portland cement per yard, occasionally by the aggregate size, and sometimes by the water content, although this characteristic may be specified only as "a little wet" or "a little dry," depending on the amount of rain that fell on the excavation the previous day.

A five-bag mix is minimal. A six-bag mix should be specified for high strength, and where reinforcing rods are placed. Aggregate should be well graded (uniform), and not more than 1½ inches in size. Aggregate 2 inches and larger may produce poor results in slab pours and other thin sections. When aggregate size is reduced from the standard 1½-1¼ inches, the cement ratio should be increased: for 1-inch aggregate add ¼ bag of cement to a five-bag mix; for ¾-inch aggregate increase the cement ratio by ½ bag; for ⅜-inch aggregate, add a full bag.

Unless the sand used is extremely dry, water content should not exceed 7½ gallons per bag of cement. For example, a typical mix with 1-inch aggregate would contain 5.8 bags of cement per cubic yard, 5 gallons of water per bag of cement, with a cement to fine aggregate (sand), to coarse aggregate (gravel) ratio of 1 to 2½ to 3½.

Outside the province of ready-mix concrete the following proportions are general guidelines for mixing concrete for footings: 1 part Portland cement, 2¾ parts sand, 4 parts gravel, and 5½ gallons of water per bag of cement. There is no mix, however, that may be followed regardless of environmental conditions, even though concrete can be protected against rapid drying in extreme heat and against freezing in extreme cold without altering the proportions of the principal ingredients.

Admixtures and Weather Protection

Technically, an admixture is any substance other than cement, sand, aggregate, and water included in the concrete to modify the characteristics of a standard mix. This group of additives includes carbon black and other coloring agents, chemicals that neutralize reaction between concrete ingredients and naturally alkali aggregate, and proprietary agents that increase bond or tensile strength in special installations. For residential applications, however, only three types of admixtures require investigation: air-entraining agents, accelerators, and retarders.

Air-entrained concrete (not to be confused with aerated concrete, which is quite different) is commonly used in road building. It contains approximately 5 percent trapped air, and is, therefore, less dense than ordinary concrete but less susceptible to frost action. The trade-off for frost resistance is roughly a 5 percent decrease in strength per 1 percent of air entrained, although this applies more to rich, high-strength mixes than to lean, lower-strength mixes. In the latter case improved concrete workability caused by the change in density generally permits a reduction in water and sand content, and this largely offsets the 5 percent strength reduction.

Air entraining also reduces the segregation of ingredients during handling, a desirable feature that maintains uniformity. Air-entrained concrete is specified by the ACI Code for all work subject to freezing temperatures when wet. The following tables detail how the compressive strength of concrete is altered by water content in air-entrained as opposed to standard concrete, and the volume of air content required for freeze-thaw protection according to the size of coarse aggregate in the mix.

The ACI Code further specifies that in order to maintain freeze-thaw protection, the water-cement ratio by weight should not exceed .53.

Accelerators, the second common type of admixture, are used to hasten the hardening rate of concrete, specifically, to reduce curing time by developing the equivalent of 28-day strengths in approximately 7 days. Proprietary agents are available, although the common admixture accelerator is calcium chloride

Water-Cement Ratios and Resulting Concrete Strengths

Compressive Strength (in psi)	Standard-Mix Concrete	Air-Entrained Concrete
	Gallons of water	Bags of Cement
2,500	7.3	6.1
3,000	6.6	5.2
3,500	5.8	4.5
4,000	5.0	4.0
4,500	4.3	3.4

($CaCl_2$) used at a maximum of 1 to 3 percent by weight of cement. Any claims that accelerators contain antifreeze, waterproofing, or hardening characteristics should be ignored.

An opposite effect is obtained with retarders. Their purpose is to provide a controlled delay of the concrete set in extreme hot weather that could cause abnormal dehydration of the mix, excessive shrinking, and a loss in strength.

However, jiggling the mix with additives is necessary only in extreme conditions. If you feel the need to experiment with proprietary additives that claim to waterproof a mix, make it crack-resistant, or generally enhance the concrete quality, do so on a wheelbarrow full, not on a foundation-sized batch, and never on only one batch in a multibatch pour.

Footing Construction and Concrete Fundamentals

Air Entrainment for Freeze-Thaw Protection

Maximum Aggregate (inches)	Air Content (Percent of total volume)
3/8	6–10
1/2	5–9
3/4	4–8
1	3.5–6.5
1 1/2	3–6
2	2.5–5.5
3	1.5–4.5

Concrete Ordering

If you mix by hand (a wheelbarrow-sized load will be less than 3 cubic feet), and the job requires only one cubic yard (enough concrete to fill less than a 30-foot run of 16 × 8-inch footing), be prepared to mix and place 9 wheelbarrows of concrete without interruption. A power mixer will speed up this process, but the rental charge, hassle of transportation, storage of ingredients on the site (protected from the weather), as well as the hard labor of mixing and placing the concrete, when considered together, make the ready-mix alternative seem well worth the extra money.

With this method of delivery, any amount of concrete can be ordered, assuring continuous pouring and a uniform mix. However, don't overdo the order. Excess concrete is needlessly expensive and difficult to dispose of. To calculate your order, apply these basic formulas and relationships: A cubic yard of concrete (referred to simply as a yard), equals 27 cubic feet (3 × 3 × 3 feet); 3/4 of a yard equals 20 1/4 cubic feet; 1/2 a yard equals 13.5 cubic feet; and 1/4 yard equals 6.75 cubic feet.

The trick is to measure the form area carefully. In fact, give yourself a safety margin by measuring it twice. For instance, if the plan calls for 10 concrete piers, each measuring 1 × 1 × 4 feet, convert the formed space to cubic volume (4 cubic feet per pier, or 40 cubic feet total), then divide by 27 (the number of cubic feet in a cubic yard), to determine the number of yards to order—in this case, 40 divided by 27 equals 1.48 or about 1 1/2 yards of concrete. To simplify calculations you can round off the cubic yard conversion factor from 27 (actual), to 25. This reduction builds in an allowance for waste, spillage, or lean calculation of about 8 percent. This is not an extravagant safety margin (plywood sheathing waste is typically figured at 15 percent), and is protection against the disaster of leaving forms partially filled until another load is scrounged up.

If form volume calculations are made in inches you must convert cubic inches to cubic feet before using the 27 or 25 conversion factor for cubic yards. For example, suppose you are calculating volume for a conventional 16-inch wide, 8-inch deep footing. For every linear foot, volume is 16 × 8 × 12 inches, or 1,536 cubic inches. Over addition or house-sized footings, however, calculations in this form will involve very large numbers, and even though elementary multiplication and division are the only tools needed to compute an accurate concrete order, almost everyone is more likely to make a silly math error as the size of the numbers increases. So instead of multiplying 1,536 cubic inches by, say, 1,440 linear inches (120 linear feet of footing), and dividing by 1,728 (the number of cubic inches in one cubic foot), simplify the math by converting to cubic feet before multiplying by linear feet.

On 120 linear feet of 16 × 8-inch footing, you can convert the volume for a single linear foot (1,536 cubic inches) to cubic feet two ways: first, by dividing 1,536 by 1,728; second, by dividing 128 (the 16 × 8-inch area of the footing) by 144 (the 12 × 12-inch area of a square foot). In both cases the result is a conversion factor of .888. In other words, each linear foot of 16 × 8-inch footing form represents .888 (.89 is close enough) cubic feet of concrete. With this method, multiplying the result by total linear feet gives a figure in cubic feet that may be directly converted to yards of concrete required.

Again, make two sets of calculations using width times depth times length the first time, then reversing the sequence of operations. A 2 × 3 × 4-foot box totals 24 cubic feet regardless of the multiplication sequence. So if two sequences give two different results, try again.

Concrete Placement

Concrete should be poured continuously and kept relatively level throughout the formed area. If it is loaded up in one corner, then raked along the form, the mix is likely to segregate because some ingredients are pulled by the rake (or shovel or hoe) more than others. This destroys mix uniformity, and may create weak spots and surface areas prone to water deterioration.

Whether you mix by hand or power mixer or order ready-mix, take a little time to set up the job site to aid continuous placement. Ideally, the swivel trough on the ready-mix truck can be positioned directly over the forms. Most trucks have a 10- or 12-foot primary chute and a 12-foot extension. But don't expect the driver to bring the truck across temporary culverts, or areas of fresh fill, or even across a cleared site that has soaked up several days of rain. If concrete must be hauled to the excavation by wheelbarrow, lay scaffold planks on the ground to provide better support and less friction against the wheel.

Concrete placed in relatively shallow footing forms (as opposed to 8-foot-high poured foundation walls) does not require much handling. A flat shovel or trowel should be used with vertical strokes along the sides of the form boards to eliminate air pockets and ensure that the formed space is completely filled. But excessive handling, in transporting, placing, or troweling off the finished form, can cause mix segregation. When the forms are filled, a short length of 2 × 4, narrow edge down, resting on top of the form boards, can be used with a gentle side-to-side motion to level and finish the footing surface (a process called screeding).

Figure 14. Independent footing and pin for wood post connection

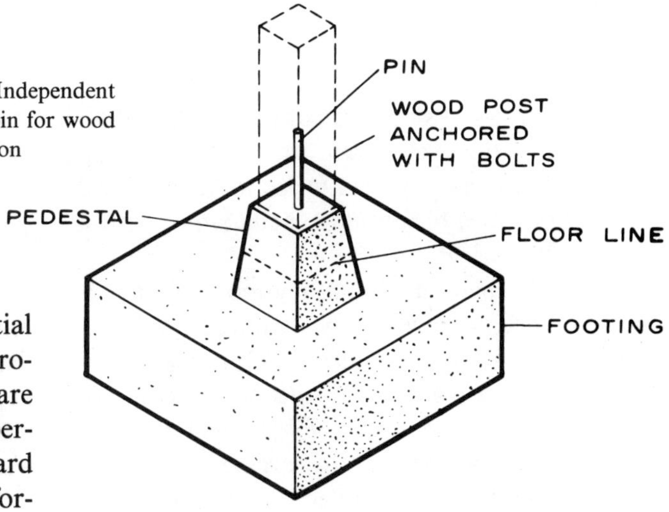

Again, a careful and accurate initial layout saves time as construction proceeds. If the 2 × 8-inch form boards are level and rigidly braced, they provide perfect guides for the 2 × 4 screed board and make finishing an easy, straightforward process.

Tying

It is surprising how often foundation walls, both poured and unit-built, are simply plunked down on the footing with only a thin bed of mortar at the joint. If foundation walls carried only vertical, compressive loads, this practice would be satisfactory. But stresses in house structures do not behave so methodically. Angled roof rafters apply a lateral thrust to side walls, while hydrostatic pressure may exert reverse lateral loads belowground. After all the time and attention devoted to footing strength, it would seem inefficient, at best, to trust this crucial construction seam to a layer of mortar.

Provision for structurally tying the footing and foundation wall together, even if poured walls are provided with a lug or keyway in the footing, can be made by embedding steel dowels (Number 4 rebars), in the footing at 12-inch intervals, extending at least 12 inches into the foundation wall. If weather or a delay between footing and foundation construction causes rust formations on the steel, use a stiff wire brush—a wire wheel bit in an electric drill is faster—to clean the bars before the foundation is built. If the wall is to be constructed of 8 × 8 × 16-inch concrete block, align the rebars to fall in the block voids.

Secondary Footings

It makes sense to pour all footings at the same time. Generally, secondary or independent footings for posts and column supports can be formed to rough elevations because the posts themselves (and many steel lally columns) can vary in height. Fireplace footings tend to be an integral part of the perimeter footing, but the excessive masonry load they carry should be independently computed.

In order to save money, many framing plans reduce floor joist size, which reduces span limits, which necessitates a central girder where the joists are spliced or overlapped. In this case a substantial load is carried in the girder span, and down the column supports to the cellar floor. Most slabs are not strong enough to carry this weight and must be backed up with piers that conform to the general principles of footing construction, with a few distinctions.

Fireplace and chimney footings, for instance, are generally required to be at least 12 inches deep, and to extend 6 inches beyond each side of the column, while conventional 8 × 8-inch piers should be adequate to carry comparatively light porch and deck loads. Piers outside the weather-protected foundation that will carry wood columns (a 4 × 4-inch post to a deck girder, for instance), should be formed to extend 3 to 4 inches above finished grade to keep the post end grain clear of moist soil that can foster rot.

Provide for a secure masonry-to-wood joint by embedding a rebar (a foot of ½-inch diameter threaded rod will suffice) in the center of the pier, projecting vertically 2 inches above the finished pour (Figure 14). To assure vertical alignment, the bar should be tie-wired to the form until the concrete sets up. When the post is erected, a corresponding hole in its base aligns with the bar projection to prevent slippage or lateral failure. (See also pages 191–93 on posts and girders.)

Stepped Footings

For the sake of economy, builders generally lay the entire footing at one level. On sloping sites this may require moving extra earth, but the big jobs of building forms and pouring concrete will be straightforward. However, you can customize your design and capitalize on a unique site by splitting the ground floor into a full basement on the low side and a crawlspace on the high side, or maintain full basement height under a multilevel first floor (a house with a sunken living room, for instance), by using stepped footings.

This design follows the general procedures of footing work within these guidelines: 1) levels must be stepped down by increments no greater than the thickness of the footing, and 2) the horizontal length of each step must be no greater than three times the footing thickness. As a general rule of thumb, and using footing thickness as a measuring unit, move one

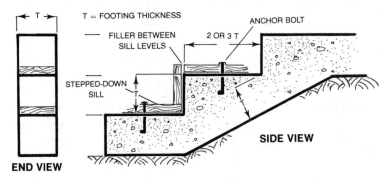

Figure 15. To accommodate sloping sites or multilevel first floors, stepped footings can be used within these guidelines.

unit down and two units forward to step down the footing.

Soil under this slope must be carefully compacted, and for additional strength rebars can be bent and laid along the slope, extending two or more feet beyond it at each end. Take extra care with concrete placement, working a trowel blade up and down along each stepped form to assure sound, clean concrete edges. Finally, it is wise to insert an anchor bolt into each horizontal step to secure the short sections of sill, which, along with its insulation and termite shield, must be continuous.

CURING AND FINAL CONCRETE STRENGTH

First, the bad news. Inadequate or improper curing of concrete can decrease strength, promote dusting, chalky surfaces, crazing, cracking, and water leakage. The good news is that proper curing is a natural process inherent in the combination of concrete ingredients and requires only isolation to run its course successfully.

Curing is the strengthening process that changes a fluid combination of cement, fine and coarse aggregate, and water into an exceptionally strong, verminproof, fireproof, long-lasting building material. Specifically, it is the chemical reaction of cement and water that occurs slowly and continuously. This hydration process is most rapid and most crucial in the first few days, and brings concrete close to its ultimate strength in 28 days.

Age to Strength Relations for Type I *(General Purpose)* **Cement**

Age *(in days)*	Compressive Strength *(in psi)*
3	2,100
7	3,200
28	4,500
90	5,000
365	5,600

Initially, concrete contains more water than required for hydration of the cement. But it is important to maintain continuous contact between water and cement particles. In combination, the by-products of hydration fill the water volume with gelatinous filaments that bind the aggregate particles together. Without enough water the migration of binding paste will be incomplete and unevenly distributed.

Several methods may be used to accomplish complete curing. The most common are moist curing and membrane curing. For Type I cement (the most widely used, general purpose variety), moist curing should continue for 14 days. During this time the concrete must remain moist from spraying, ponding, or by covering with burlap that is periodically saturated with water. This is the method most frequently illustrated in do-it-yourself books, although most do not point out its disadvantages, namely, your almost constant presence at the site. If the sun comes up at 6 o'clock A.M., the burlap may well be completely dry in an hour. If the evening wind is warm, it may be dry hours earlier.

Membrane curing, the more practical and widely used method, is achieved by covering and sealing the concrete with a layer of impermeable polyethylene sheeting, at least 4 mils thick. Alternatively, a uniformly thick film of curing compound can be sprayed on the surface, although this method is more practical on large-scale slabs than on strip or pier footings. Coverage standards are 150 square feet per gallon for wax-base compounds, and 200 square feet per gallon for resin-base compounds. Both may be applied with a conventional paint sprayer, but an experienced eye is needed to assure uniform coverage. Film thinning, particularly around edges and corners, can lead to significant strength loss from accelerated hydration, and moisture loss by evaporation. Remember, too much water makes the mix lean and decreases strength, while not enough water limits complete hydration with the same result. Both types of membrane curing (over 28 days) are roughly equivalent to 14 days of continuous moist curing.

Over 28 days, high strength is achieved with a good, uniform curing agent. By comparison, low-grade or intermittent curing will produce about 65 percent of 28-day strength, while the absence of curing control can reduce 28-day strength by 50 percent. In all cases the first 3 days are the most important.

The relationship of cement and water in concrete is crucial in curing, and in attaining strength, which increases as the water to cement ratio decreases. Additionally, less water in the mix (only up to the point where complete hydration is still possible), makes the binding paste denser, less permeable, and, therefore, more watertight.

Footings occupy the space between the house and the ground. They take the weight of every 2×4, every nail, every shingle. If they fail, everything above them fails. The best policy is to remove any reasonable doubt about layout accuracy, frost depth, soil type, and other crucial details before ordering concrete. Even then, at the finishing stage, when thorough planning and workmanship are about to pay off, take the time to consider mix ratios and reinforcement.

The step-by-step process of house construction has many characteristics of a stack of toy blocks. If the first one rests on a flat floor or table, the potential for a high, stable stack of blocks is preserved. If the second block is placed just a little out of line, the potential is reduced. Successive errors, even marginal misalignments, compound the chance of failure until placing one block on another becomes an adventure.

Building is interesting enough without this kind of excitement. Footings are the first step. They set the pattern and the quality of the job. And careful attention to footing design and construction preserves the potential for building a solid, minimal-maintenance, long-lasting home.

5

FOUNDATION CONSTRUCTION AND BLOCK FUNDAMENTALS

The worst is behind you. That's right. The excavation work, the planning, the layout, the footings, all the literally ground-breaking parts of construction are done with. Now, from the footings up, you begin to build with modular materials, to enclose space, to give form to the blueprints.

Of course, there is still room for error, and even the most experienced, conscientious general contractor will run into problems. Some are inevitable. But the trick is to minimize the errors and to preserve, as nearly as possible, the potential for a sound, long-lasting structure through each successive stage of construction.

Initial planning, the layout, and the footing construction are formative steps that give purposeful dimension to a bare site. Following through on the design is now largely a matter of execution, using one of several foundation materials.

Masonry is the most common foundation wall material, although advances in wood preservative technology have opened up another alternative, wooden foundations laid directly on the ground. Masonry foundations can be built all at once, with poured concrete, or piece by piece, with concrete block. Each material has advantages and disadvantages that vary according to the building design, characteristics of the site, local labor and material rates, weather conditions, and other factors. In other words, one is not obviously more desirable than the other.

Labor on poured walls is concentrated on formwork. Thorough construction requires that reinforcement, anchors, ties, provision for utilities, a soil pipe, pockets for girders, vents, and window openings must all be built into the form. This process is really 90 percent carpentry work. Compared to the time expended to transfer every angle and shape to the form, and to brace them sturdily enough to withstand the lateral loads and pouring force of liquid concrete, pouring the walls is only the finishing touch.

On additions or house-sized projects, you should rent prefabricated forms or hire a subcontractor who will provide them. You can, of course, make your own, and you should for individual piers or steps, for instance. But house foundations and additions require a lot of plywood and a lot of 2 × 4's. (Remember, two wall faces must be formed for each side of the foundation.) And most of this lumber cannot be reused in the superstructure framing once the forms are stripped. Such a large, one-time expense should be avoided.

Generally, modular forms that clip together, that have provisions for wall-to-wall ties, and that can be used with components to shape foundation vents and windows, can be erected in less time than it takes to build the same walls out of concrete block. That is the beauty of modular construction—you can form the wall to its full height in 4-foot sections. However, the concrete walls must be left undisturbed for at least two days, preferably longer, even in relatively dry, warm weather. In cold weather you might have to wait a week or more before the concrete has acquired enough strength to support the loads of further construction. And as long as the forms are in place floor framing cannot be laid.

On the other hand, if it makes sense to build footings of solid, reinforced concrete instead of individual concrete blocks, shouldn't the same thing be true for foundations? Certainly, poured walls are monolithic and unlikely to crack if properly reinforced and provided with control joints. Poured at one time along with the footings, they can become a single, immovable object that resists small pockets of abnormal soil settling or erosion.

Furthermore, foundation walls can be made with a concrete mix richly proportioned to provide excellent water resistance. And there are other advantages. Colorant can be added to the mix and textures applied to the surface. Concrete block, on the other hand, looks the part of the job it does, and, to suit most tastes, must be stuccoed or covered in some other way to hide the joints. (Oddly, most people do not have this reaction to the joints between bricks.)

Poured concrete walls can be finished in a variety of textures as they are poured. This saves one stage of construction and the inevitable maintenance that results when different materials are joined over a large surface area. A poured wall reflects the image of the form surrounding it. For every formed protrusion ($\frac{1}{4}$-round strips could be nailed to the form face in 4-inch horizontal runs, for example), there is a concrete depression. A finished, above-grade foundation wall can be vertically or horizontally striated, imprinted to simulate textured brick, the swirls of troweled-on stucco, or even the grain of rough-sawn plywood.

The National Ready-Mixed Concrete Association points out that Civil Defense groups recommend concrete basements for refuge during storms or nuclear attack, and that concrete basements can be modified to make holding chambers for forced-air, solar-system rock storage—the old and the new. Additionally, although still experimental, concrete is being used for massive, direct-gain foundations, the principle being that below the frost line, soil holds sufficient heat to warm a basement or larger area, provided enough concrete is placed to absorb and transfer the latent heat.

House foundations can be successfully built with concrete block as well. They are available in a wide variety of sizes and configurations, just like toy building blocks, so you can build vertical walls, sills, window openings, pilasters to support girders and beams, interlocking control joints, cavity walls, you name it. Forms are not required, but as successive courses of block are laid, they must be aligned, leveled, and plumbed.

In residential construction, block size is conventionally 8 × 8 × 16 inches (nominal), which actually measures $7\frac{5}{8}$ inches wide and $15\frac{5}{8}$ inches long to allow for $\frac{3}{8}$-inch mortar joints. In this way, finished blocks run in multiples of 16 inches center-to-center, which matches the modular layout for 2 × 4 (and the newer 2 × 6) stud walls. The blocks most commonly used are three-core stretchers (8 × 8 × 16 inches with U-shaped extensions at each end to form half a common core with adjoining block); three-core single corners (8 × 8 × 16 inches with one square-faced end to be exposed in the exterior wall face); and solids (4 × 8 × 16 inches with all sides square-faced) to serve as capping.

While poured walls have no joints other than built-in control joints to relieve surface tension stresses, block walls, which need reinforcement and control joints as well, are littered with joints. In almost every case they are assembled in a staggered pattern (called a running bond), although vertical and horizontal joint alignment (called a stack bond) may be used with additional reinforcement.

The keys to a good poured concrete foundation wall are true, rigid forms, and a uniform, high-strength concrete mix. For block construction the keys are constant and consistent alignment, block after block, and a uniform mortar mix with high bonding strength. The final strength and durability of both materials can be significantly improved with tying and reinforcement, which requires roughly equal effort in each case.

The most basic distinction between the two materials, though, is in the process of construction. With poured concrete, details and accuracy are built into the layout and the forms. The concrete simply fills the space provided for it. With block walls, construction is drawn out into pieces, and each one must be attended to individually.

BLOCK FUNDAMENTALS

Standard concrete blocks used in residential foundation walls are only one type of concrete masonry unit, a category of building materials that includes concrete building brick, tile, slab sections for floors and roofs, lintels, and other configurations. Concrete block itself can be made with sand, gravel, crushed stone, coal cinders, expanded shale, pumice, or other aggregates bound with Portland cement. Each material imparts different properties to the block, but as a rule the term concrete block refers only to masonry units made with sand and gravel or crushed stone aggregates.

Unlike poured concrete, the blocks are fully cured by the manufacturer, although methods vary widely, from simply storing the block in an exposed area to sophisticated, high-pressure steam curing. And these diverse methods can make a significant difference in block strength and durability. For instance, field tests have shown $\frac{1}{4}$- to $\frac{3}{8}$-inch shrinkage per 100 feet in steam-cured block, and roughly twice that much in atmospherically cured block—enough to encourage cracking.

Structural specifications, determined by the American Society for Testing and Materials (ASTM), for hollow, load-bearing block include parameters for wall and web thickness, and other characteristics. However, for computing structural loads, two figures are important. For block graded Type N and Type I (moisture-controlled block for general use above or below grade), individual compressive strength is 800 psi (pounds per square inch). For block graded Type S-I (moisture-controlled block used above grade with weather-protective coatings), minimum, individual compressive strength is 700 psi.

Concrete block is abrasive and heavy, averaging about 55 pounds per square foot of wall area. For 8-inch thick walls made of general-use, sand-and-gravel aggregate block ($8 \times 8 \times 16$ inches nominal), 100 square feet of wall area would create a dead load of 5,550 pounds on the footing. This area would require 112.5 blocks, and an average of 8.5 cubic feet of mortar to form the joints.

In addition to strength and durability, block walls provide excellent fire resistance. Attached houses, constructed largely of minimal, highly insulated wood frames, are frequently joined at a common wall of concrete block, constructed across the full cross section of the buildings to form a fire stop.

In areas where fire resistance, or, more specifically, a material's structural endurance under fire conditions, is particularly important, aggregate composition should be considered. This resistance is expressed in equivalent thickness (ET) for hollow block and varies according to the type of aggregate used in the block. The hour measurements refer to how long the material maintains its structural integrity under fire.

Different grades and types of concrete block should be specified for particular building conditions. For instance, Type I moisture-controlled block should be used where shrinkage caused by drying could result in cracking. Be sure of the grade, type, and characteristics of the block you order before combining them in a foundation wall.

Concrete Block Data

GRADES OF CONCRETE BLOCK N—For general use in exterior walls above or below grade (or interior walls) that may be exposed to moisture penetration or the weather. S—For limited use above grade only with protective coatings, and for walls not exposed to weather.

STRENGTH OF CONCRETE BLOCK Hollow, N-grade block (designated C90 by ASTM)—800 psi compressive strength per unit, 1,000 psi averaged over three units. Hollow, S-grade block (same designation)—600 psi compressive strength per unit, 700 psi averaged over three units. Solid, N-grade block (designated C145 by ASTM)—1,500 psi compressive strength per unit, 1,800 averaged over three units. Solid, S-grade block (same designation)—1,000 psi compressive strength per unit, 1,200 averaged over three units.

Mortar for Concrete Block

The mortar in block walls must be capable of performing several functions. First, it must join individual blocks together to form a single structural element. At the same time, these joints must remain sealed to air and moisture, even when the mortar is in contact with steel reinforcement, anchor bolts, or metal ties.

Mortar should not be confused with grout, which is used predominantly with reinforced construction. Grout is placed in the walls to bond the masonry units and reinforcing rods together, and sometimes to give added strength to a block wall that is not reinforced.

Mortar is available in dry-mix bags, or you can mix it yourself. Two types, both with high compressive strength, are used in most residential applications, including solid and hollow unit foundations and piers. Type M mortar is proportioned by volume of one part Portland cement, $\frac{1}{4}$ part hydrated lime, and $2\frac{1}{4}$ to 3 parts sand in a damp, loose condition. It reaches a 28-day compressive strength of 2,500 psi. Type S mortar is proportioned by volume of one part Portland cement, $\frac{1}{2}$ part hydrated lime, and $2\frac{1}{4}$ to 3 parts sand for a 28-day strength of 1,800 psi.

However, these mortars greatly exceed the compressive strength of ordinary block, and because of their low lime content, are difficult to work with. A typical, very workable mortar mix for concrete block subjected only to normal service would be a ratio of one to one with Portland cement and either hydrated lime or lime putty, combined with 5 to 6 parts of sand.

As with concrete, admixtures may be combined with the mortar to reduce early water loss by suction to the dry concrete block, to promote bonding, and to produce slight expansion so the mix fills all cavities. However, calcium chloride should not be used, particularly in grout, because of its corrosive action on reinforcement. On exposed walls white mortar can be made using white masonry cement; or it can be made with a combination of white Portland cement, lime, and white sand. And color pigments may be combined with the dry mix to produce a variety of shades.

On most jobs, with the exception of very small ones, mortar should be machine mixed. Portable machine mixers can produce between 4 and 7 cubic feet per load (8.5 cubic feet is needed for 100 square feet of wall). Good results can be obtained by batching roughly ¾ of the water, ½ the sand, and all the cement and mixing briefly, then adding the remaining ingredients for a total mix time of three to five minutes. Less mixing time can create a lumpy, nonuniform mix. Mixing too long can adversely affect air-entraining agents and can reduce mortar strength.

If you do mix the mortar by hand, combine the dry ingredients thoroughly (even if they come out of a premixed bag). Then add about ¾ of the water, and continue mixing until the mortar is uniformly wet. Finally add the remaining water sparingly, mixing thoroughly to achieve proper workability. Each batch should be allowed to stand about five minutes and then remixed with a hoe before it is used.

Water content is generally 4 to 5 gallons per cubic foot, although the condition of the sand and the concrete block can affect moisture content. You can test for proper water content by cutting a curved furrow in the mortar with a hoe. If the sidewalls of the furrow hold their shape, and mortar can be cleared from the hoe with a brisk shake, the mix is properly balanced. If the walls of the furrow collapse, however, too much water has already been used, and small amounts of premixed, dry ingredients must be added and thoroughly spread into the mix. If the mortar cannot be shaken off the hoe, the mix is too dry, and small quantities of water should be added until the mix becomes workable.

Obviously, it is a lot easier to mix mortar than it is to lay block and certainly a lot quicker, with the mixer running only four or five minutes. Consequently, a lot of the mortar may sit around before being used. Mortar that is stiffened by hydration should be discarded. But mortar that has stiffened because of evaporation can be brought back to a usable condition by retempering, that is, by adding water as needed and thoroughly remixing.

Given the small difference between hydration, or actual setting of the mortar, and simple evaporation, it is generally safer to follow these guidelines. Use mortar within 2½ hours of mixing when the temperature is 80°F or above, within 3½ hours below this temperature; discard it after 3½ to 4 hours maximum.

Design Alternatives with Concrete Block

Concrete block can be made with different types of aggregate; it can be made into 2- or 3-core units, with smooth, split, or ribbed faces; reinforced; joined with different types of mortar—there are many decisions to make. Additionally, even the most basic 8 × 8 × 16-inch foundation wall can be altered by varying the bond pattern.

The most conventional running bond places the joints between blocks in one course directly over the midway point of each block in the course below. But there are many other alternatives. For instance, joints can be staggered 4 inches to the left and right of preceding joints to create an offset stagger, or simply offset 4 inches in one direction through each course. And by including a second type of dimensional block, alternate courses of $4 \times 8 \times 16$-inch block, for instance, a large number of wall patterns are possible.

CONCRETE BLOCK CONSTRUCTION

After design decisions have been made, the next step, really the first step in construction, is to determine the amounts of materials needed for the job. Use the following tables for material estimates, with a reminder that mortar quantities include a 10 percent waste allowance, which may be less than some contractors allow.

For reinforced block walls where the voids are filled with grout (proportioned at 1 part Portland cement, $1/10$ part lime, and $2\frac{1}{4}$ to 3 parts aggregate), the following quantities are necessary, and include a 3 percent waste allowance.

If you mix the mortar at the site from raw materials, first determine from the preceding tables how much you will need according to the number of blocks required for the job, then use the following table to determine the amount of each mortar ingredient required.

Layout

Sweep the footing surface clean, and establish a work area near the wall where you can stack block and mix mortar without tripping over tools and materials. And whether the mortar is machine- or hand-mixed, set up a mortar table (simply a piece of plywood resting on several blocks), right next to the wall (Figure 16). Once you start working, it is encouraging to see the wall take shape, and discourag-

Estimating Data for Concrete Block

Wall Thickness (inches)	Block Size (inches)	Number of Blocks (per 100 square feet wall area)	Cubic Feet of Mortar (per 100 square feet wall area)	Average Weight (pounds)
6	$6 \times 4 \times 16$	225	13.5	5,100
6	$6 \times 8 \times 16$	112.5	8.5	4,600
8	$8 \times 4 \times 16$	225	13.5	6,000
8	$8 \times 8 \times 16$	112.5	8.5	5,550
12	$12 \times 8 \times 16$	112.5	8.5	7,550

ing to stop every few minutes to carry a few blocks to the wall or bring over another small load of mortar.

First, though, you should go back to the batter boards, restring any lines that were removed during footing construction, and check to assure yourself that the boards, and in turn, the layout, have not been disturbed. Then transfer the building lines to the footing surface, paying particular attention to the corners. This can be done in several ways.

If you have extreme confidence in the footing layout (remember, it will extend beyond each side of the block wall), you could simply measure in to find the center line of the footing, and, if 8-inch-wide block is being used, mark the inside and outside of the block wall at 4 inches from the center line. If you do proceed this way use a chalk line (string wound in a box of colored powder that, when stretched between two points and snapped, leaves a colored line) to mark the perimeter of the foundation. Then measure the walls and the diagonals to assure accuracy and squareness.

If, however, you feel more comfortable measuring off the original layout lines, the initial points and corners painstakingly laid out with the batter boards, you should make sure that the lines between the boards are taut, then transfer the corner intersections to the footing. Many trade manuals suggest that you accomplish this by attaching a plumb bob on a string to the string intersection above the wall. But accurate plumb bobs are heavy, and may distort the plumb transfer by pulling on the line unevenly.

This distortion can be eliminated by es-

Estimating Data for Grout

Wall Thickness (inches)	Spacing of Grouted Cores (inches)	Cubic Yards of Grout (per 100 square feet wall area)
6	all grouted	0.79
6	16	0.40
6	24	0.28
6	32	0.22
6	40	0.19
6	48	0.17
8	all grouted	1.26
8	16	0.74
8	24	0.58
8	32	0.49
8	40	0.44
8	48	0.39

Estimating Data for Mortar Ingredients

Mortar Mix Proportions by Volume/Quantities per Cubic Foot of Mortar

Type	Portland Cement	Masonry Cement	Hydrated Lime	Sand
M	1/0.16	1/0.16	—	6/0.97
M or S	1/0.29	—	1/4/0.07	3/0.96

NOTE: aggregate should not be less than 2¼ and not more than 3 times the sum of the volumes of the cements and limes used.

Foundation Construction and Block Fundamentals

tablishing a plumb line, about an inch away from the string intersection on each side, with a 4-foot level. Slight adjustments can be made to find a plumb line as the level just barely contacts the string. When these marks are set on the footing surface, extend them until they intersect, repeat the procedure at each corner, then use these points as the ends of a chalk line to establish a reliable foundation wall perimeter. And as a practical safeguard, make a series of marks about 2 inches outside these lines. When the first course of block is laid (try a dry run first, Figure 17) the full mortar bed is likely to obscure the line temporarily, but with the 2-inch benchmarks, block alignment can be checked in any case.

Mortaring Technique

The idea is to get a uniform amount of mortar on the horizontal and vertical edges of the block to assure complete bearing without losing too much mortar. If the trowelful falls onto the ground you should throw it away. Although ten professional masons will have ten different styles of applying mortar, you should get good results with these guidelines.

Apply mortar to horizontal joints by cutting a slice of mortar from the pile the way you might scoop up a piece of pie from a pan. This slice of mortar, called a windrow, should be concentrated on one side of the trowel. Deposit it on the block with a dropping motion, sharply halted just as the trowel reaches the block. It makes sense to work from a position where the face of the trowel will wind up

Figure 16. Tempering mix from a centrally located mortar table

Figure 17. Dry run of block layout to perimeter chalk line on first course

Figure 18. Furrowing the first-course mortar bed

Figure 19. Buttering the ends of first-course block

Figure 20. Buttered end is laid down and moved in to preceding block.

against the outside edge of the block. This way, if you miss or throw the mortar too vigorously, the blade will stop the mortar from falling off the block. When enough mortar has been placed for three or four blocks, spread it out using the trowel in an inverted position, and dragging the point down the center line of the mortar (Figure 18). With practice, this process, called furrowing, can be used to spread an even layer across several blocks.

When the horizontal mortar beds are ready stand several blocks on end (three is a convenient amount because with the mortar joints, total length should equal 4 feet, the length of a level), and apply mortar to the faces of the shell extension (the surfaces where blocks will bear against each other). This process, called buttering, should maintain the $\frac{3}{8}$-inch joint spacing (Figure 19).

In all cases, somewhat more mortar than the final $\frac{3}{8}$-inch amount must be placed. Then, as the block is set in place, and tapped into horizontal and vertical alignment (generally with the heel of the trowel), this excess mortar will ooze out of the head and bed joints. Properly proportioned and mixed mortar, however, should hold its shape so that after the block is set it can be cut off flush with the block surface, using the edge of the trowel.

First Course Construction

On the first course of block, full mortar bedding should be used. Therefore, enough mortar must be placed to establish continuous bearing between the edges of the face shell, the edges of the webs that separate the block voids, and the footing. On the first course, however, it is wise to put the blocks in place dry, as a test run, including spaces for the $3/8$-inch mortar joints, to determine if full blocks can be run corner to corner (this requires an exceptionally well-planned and executed layout), or, if not, where a filler block or half block will be cut. (See Figure 17, page 69.)

As each block is set in place, starting at the corners, tap it into position firmly, aligning each one with the perimeter chalk line and with adjacent blocks. To do this, use a 4-foot level, checking along the face and top of the block at regular intervals. Remember that following courses will be aligned with this first layer of blocks. If not enough mortar is placed, pick up the block, apply more, and reset it. For continuous bearing, mortar should protrude from the joints as the block is plumbed and leveled.

Figure 22. Height is corrected by tapping with heel of trowel.

Figure 23. Checking plumb alignment of first-course corner block

Figure 21. Block is adjusted and set on the full mortar bed.

Figure 24. Leveling and checking edge alignment of first-course block

FOOTINGS AND FOUNDATIONS

Figure 25. Built-up, interlocking corner

Figure 26. Checking alignment of mortar joints

Figure 27. Plumbing built-up corner

Corner Construction

It is customary to pay particular attention to the corners, building them up 4 or 5 courses higher than the runs of block between them (Figure 25). Obviously, you should place all blocks accurately, but with this method the corners, in effect, act as horizontal and vertical controls as they represent the original intersection of the batter board lines.

Corners should be interlocked, that is, the 8-inch face of one course should be covered by the 16-inch face of the block above it. Build away from the corners in each direction in a stepped pattern. Never try to cantilever a block (suspend half its length beyond the block beneath it).

In addition to checking carefully with a 4-foot level for horizontal and vertical alignment, hold the level at a 45° angle along the steps in successive courses at each corner (Figure 28). If there are minor discrepancies they must be absorbed

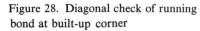

Figure 28. Diagonal check of running bond at built-up corner

Foundation Construction and Block Fundamentals

in the mortar joint. And since these joints are almost always set by hand (though mechanical spreaders that ride along the block surface can be rented), they are the most likely source of layout problems. A fractional error, only $\frac{1}{32}$ inch of thickness, compounded course after course, can significantly alter horizontal alignment.

One of the easiest ways to control height, that is, consistent thickness of mortar beds over several courses, is to lay out each new course at the corners with a story pole. To make one, lay out square lines across the edge of a straight 2 × 4 exactly 8 inches apart (the depth of the block plus $\frac{3}{8}$ inches for the mortar joint). As each course is raised, stand the story pole on the footing and against the corner. In an accurate layout the marks should be level with the top of the block, not the top of the mortar bed (Figure 31).

Figure 29. Course intervals (8-inch) checked against marked story pole

Filling In

Blocks can be laid accurately and efficiently if most of the space filled is predetermined. If the corners are built up accurately, laying the block between them is just filling a premeasured, preleveled space, and should not take long. Again, it makes more sense to go back to the original layout, to transfer the crucial corners up the wall, and generally to lay out accurate parameters than it does to measure each block.

Using the corners, you can set mason's blocks at each end of the wall with a line suspended tautly between them. This can serve as both a vertical and horizontal guide, although you should make periodic checks with a level.

While a full bed of mortar is used on the first course, it is common practice to apply mortar only to the face shell, not the webs that separate the voids (Figures 33 and 34). But even though this takes less time than spreading a full bed, all adjustments must be made before the mortar starts to stiffen. And if realignment is necessary, the original bed must be removed and fresh mortar applied to assure a good bond and watertight joints. Clean away the excess mortar as the block is laid, reusing the mortar while it is still pliable.

Closure Block

As courses are filled in from the corners, a point is reached where the final block, called the closure unit, must be fitted (Figures 35 and 36). Before fitting

Figure 30. Block is tipped into place and checked against line.

Figure 31. Adjustments for final positioning are made with trowel heel.

Figure 32. Excess mortar is cut off with trowel edge.

Figure 33. Face shell mortar bed is used after the first course.

Figure 34. Buttered block is laid on the face shell bed.

Figure 35. Buttering the edges of the closure block space

Foundation Construction and Block Fundamentals

this last block in the course, all edges of the opening and all edges of the block should be buttered. If a full block is too tight, or if the closure unit has been miscut, remove the block and start again. Think of this unit as the keystone in an arch. Certainly, it is not as crucial in a structural sense, but the last opening in the course should not be the space where all discrepancies are hidden. Also, do not stack closure units above each other in the wall. Stagger them at varying intervals in successive courses.

Cutting

Most masonry supply yards sell half sizes of block that may eliminate the need for cutting on modular layouts. But when special conditions dictate a custom fit, blocks can be cut with an overhead-blade masonry saw (Figure 37—strictly a commercial tool) or by hand.

To obtain a clean break in the block, the line of cut should be heavily scored on both faces using a brick set and hammer (Figure 38). When the line has been cut into the block, a few sharp blows on the brickset should produce a clean break.

Figure 37. A commercial contractor's masonry saw

Figure 38. Scoring the block with a brickset and hammer

Figure 36. Setting the buttered closure block

Figure 39. A ⅝-inch-diameter round bar for concave joint tooling

Figure 40. A ½-inch square bar for V-groove joint tooling

Tooling

Although mortar joints trimmed flush using the trowel edge are satisfactory, appearance and weathertightness can be improved by compacting the mortar after it has set up. The most common tooling pattern is a concave shape that may be produced with a tooling iron or even a short length of pipe.

For horizontal joints the tool should be at least 22 inches long and slightly upturned at the lead end to prevent gouging. Concave joints can be tooled with a ⅝-inch diameter pipe, while a ½-inch-square bar produces V-joints when held at an angle to the mortar (Figures 39, 40). Tool the horizontal joints first, then use an S-shaped jointer to strike the vertical seams (Figure 41). Final mortar burrs can be removed with the trowel edge or by rubbing with a piece of burlap.

Openings

When openings are made in the foundation wall, loads that would have been supported by the blocks in the opening must be transferred to the sides of the window or vent space. This can be accomplished with precast concrete lintels. Under the bearing ends of the lintels noncorroding metal plates should be laid beneath the mortar bed (Figure 42). Complete the opening control joint by raking out ¾ inch of set mortar at each vertical joint, and filling the groove with caulking (Figure 43—for more about control joints, see page 79).

Alternatively, stock-sized foundation

Foundation Construction and Block Fundamentals

Figure 41. An S-shaped jointer for striking vertical joints

Figure 42. Metal slip plate under cast lintel for joint control

Figure 43. Mortar joint is raked for cracking control.

Figure 44. Bearing wall reinforced joint with tiebar in solid voids

Figure 45. Metal lath supports mortar in void that embeds tiebar

Figure 46. A 28" tiebar locked in position

vent windows can be placed in the top course of block and covered with the wooden foundation sill. This construction is more economical, faster, and results in windows and vents well above finished grade for protection against leaking from splattering or drifted snow.

Ties

If the blueprints call for an intersection of bearing walls, for instance, where the excavated area is divided between full height cellar and crawl space, the T-joint must be reinforced. The walls should not be tied together in a masonry bond (the kind used at corners), but with straight butt joints strengthened with metal tiebars.

Tiebar is ¼ inch thick, 1¼ inches wide, and 28 inches long, with 2-inch right-angle bends in opposite directions at each end. For good joint control the tiebars should be placed no less than every sixth course with the bent ends embedded in a block void filled with mortar or concrete. This material can be supported by placing a patch of wire lath in the previous course, directly beneath the tiebar (Figures 44, 45, 46).

Control Joints

No building material is completely static. Concrete will change moisture content according to the weather. It will shrink as it dries out. Control joints are intended to concentrate these cracking stresses in a controllable, inconspicuous way.

When you drive over a large bridge, and suddenly your tires hum for a second, you have just driven over one kind of control joint—those interlocking steel teeth embedded in the roadway. One day the teeth may be closely interlocked, and the next day overlapping only a few inches. This system permits the bridge to move without disrupting the pavement.

In foundations, control joints are recommended at the following locations: 1) at abrupt changes in wall height; 2) at changes in wall thickness, for instance, where a pilaster is added to support a girder; 3) at one or both sides of window and door openings; and 4) at intervals in long, straight walls. If the wall contains no reinforcement, full height, vertical control joints should be placed every 40 feet. For 24-inch center reinforcement, they should be placed every 45 feet; for 16-inch, every 50 feet; and for 8-inch, every 60 feet.

Special interlocking block (tongue-and-groove) may be used in full or half sizes with caulking compound instead of mortar in the vertical seam (Figure 47). With conventional block, control joints can be created by raking out the vertical seam to a ¾-inch depth and filling the seam with compound (Figures 50 and 51). In many

Figure 47. Tongue-and-groove block for control joints

Figure 48. Manufactured half-block for vertical control joint

Figure 49. Control joint alignment with half- and full-block

cases the manufacturer's recommendations include priming the edges of the masonry units to prevent moisture from escaping into the porous block.

Other methods include laying a sheet of tar paper at the vertical joint between blocks to prevent mortar from bonding to blocks across the joint (Figure 52). This joint, called a Michigan type, and control joints formed with interlocking, offset jamb blocks, should be laterally reinforced by placing a Z-bar along the seam (Figures 53 and 54).

Capping

On foundation walls the top course must be solid to provide continuous bearing, and sealed to act as a secondary termite barrier. If conventional stretcher block is used, as is common on perimeter foundation walls that will support a wood frame, lay a continuous strip of metal lath underneath the top course, and fill all voids flush with the top of the block (Figures 55 and 56).

Figure 51. Control joint is caulked to provide a uniform surface.

Figure 52. Roofing felt prevents mortar bond in control joint.

Figure 50. Vertical control joint is raked out to a ¾-inch depth.

Figure 53. Interlocking joint made of reversed jamb block

Figure 54. A Z tiebar is used in every second course over control joint.

Figure 55. Full metal lath under top course supports mortar in voids.

Figure 56. All voids in the top course are filled.

Anchor bolts (½-inch diameter, 18 inches long, with a bend at the non-threaded end), which will be used to attach the sill, should be set in filled block voids along the top course not more than 4 feet apart (Figure 57). Threaded ends should protrude about 2 inches, allowing for 1½-inch thickness of wood sill and leaving enough room for a washer and nut to secure the sill to the foundation.

For masonry walls that will not need a wooden sill, capping can also be made with special solid-surface block or with solid 4-inch thick units (Figures 58 and 59).

POURED CONCRETE CONSTRUCTION

Pouring foundation walls, as opposed to laying block, is a one-shot proposition, dependent on two key elements for success—the forms and the mix.

The quality of block walls is determined most by workmanship, given the use of common, standardized building materials. At every joint the misalignment of block or skimpy use of mortar can create unsound, leak-prone chinks in the wall. But this is unlikely with poured concrete construction. If there is going to be a problem—if the mix is not protected from freezing, if the forms are inadequately braced—the mistake will register throughout the foundation.

Concrete is used in many design applications, and may be subjected to abrasion, heat, chemical attack, and severe

Figure 57. Anchor bolts to secure wooden sill are set in filled voids.

structural loads. But the demands on concrete in residential foundation walls are light, and easily within the material's performance characteristics. In this case the criteria are simple: Walls must be straight, plumb, crack-free, and of sufficient strength to support the house.

Concrete Foundation Forms

Forms for foundation walls fall into two categories, site-built and prefabricated. Generally, and particularly on large, straightforward projects, prefabricated forms are the most economical, simply because their initial cost can be spread over many jobs as the forms are reused. Site-built forms for highly customized

Foundation Construction and Block Fundamentals

Figure 58. Solid 4-inch block for capping

Figure 59. Solid-surface block for capping course

plans constitute a considerable, one-time expense. Nevertheless, when material and labor costs permit, site-built forms will produce high-quality concrete walls.
SITE-BUILT FORMS The most common site-built form materials are 4×8 sheets of $5/8$- or $3/4$-inch B-B grade exterior plywood, and construction grade 2×4's used as reinforcement and bracing. B-B grading indicates knots in the form surface, which are guaranteed not to pop out, while exterior grading adds protection against delamination of the plywood layers caused by prolonged exposure to wet concrete. Although lesser face grades detract only from the appearance of the finished concrete, exterior grades are made with water-resistant glue that permits maximum reuse.

The highest quality finishes can be produced on forms made of overlaid plywood (a conventional sheet with a resin-impregnated fiber face, fused to the sheet under heat and pressure), which leaves a gloss surface on concrete. Tempered, $1/4$-inch hardboard with a plywood backing, tongue-and-groove, shiplap, or square-edged boards may be used as form faces as well.

The principle of site-built form construction is similar to the design of a stud wall. Place 2×4's 16 inches center-to-center with horizontal 2×4's at the top and bottom of the form. Additionally, since the concrete exerts lateral loads, the vertical 2 × 4 studs are backed up with horizontal members, called wales, at 2-foot intervals. These supports are generally made of two 2 × 4's spiked together for extra strength.

For free-standing foundation walls, two form faces are required, and each one must be braced with 2 × 4's between the horizontal wales and the ground, where stakes should be driven to pin the angled braces securely. All of this construction is temporary. So in order to speed stripping and removal, use double-headed nails.

Wood forms can perform well under the loads of pouring concrete because wood in general has a capacity to absorb large loads for short periods. But even with vertical and horizontal bracing, ties, inserted wall to wall, are generally required to keep deflection in form walls below $1/240$. This means that a sheet of plywood would bow out only .07 of an inch between 16-inch supports.

Many varieties of ties are available. The idea is simply to bind the two sides of the forms together with wire, inserted through holes in the forms and wrapped around the horizontal wales. Most can be partially or completely removed from the concrete after it has set up, frequently by snapping off the retaining hardware just beneath the concrete surface. Ties can be inserted between the double 2 × 4's of each wale, no greater than 36 inches apart.

As the ties are tightened to resist the pouring force of concrete, spreaders

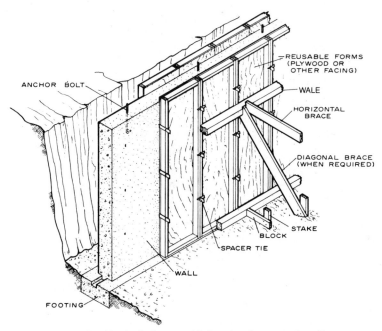

Figure 60. Site-built forms with bracing for poured walls

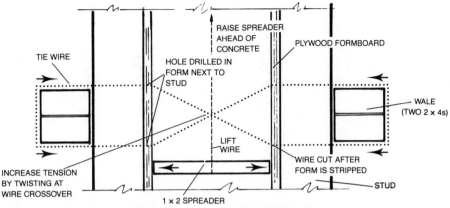
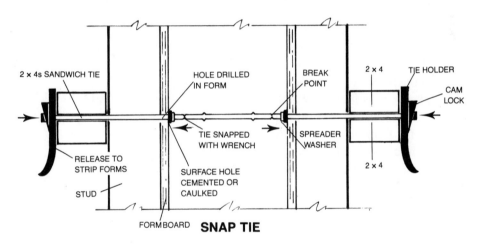
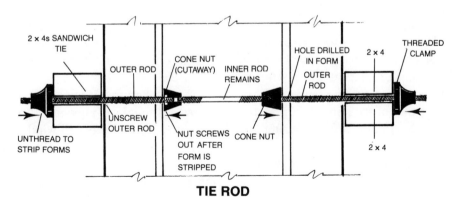

Figure 61. Form details: section views

(8-inch lengths of 2 × 2, for instance) are inserted in line with the ties to prevent the forms from buckling toward each other. Wooden spreaders should not be buried in the concrete. When they are placed, tie wire should be wrapped around each spreader in vertical alignment so that they can be dislodged before the concrete sets.

The most efficient method, however, is a spreader/tie combination, which leaves only small holes that are easily patched with cement mortar. Many manufacturers supply ties in strips that must be broken to 8-inch lengths.

Concrete is a dense, hard material, difficult to drill or cut. That's why when you are constructing the forms you should plan for openings in the foundation walls such as basement vent windows and pockets for floor beams, and holes for electrical conduit or a soil pipe.

Concrete can be stopped at these openings by using frames of ⅝ or ¾-inch plywood that run perpendicular to the form walls. To get a clean concrete edge, the frame must be cut and installed accurately. The key to planning these openings is remembering that the thickness of the material you use will also take up space, and must be accounted for in the rough opening size. For conduit and piping, the best solution is to cut the correct diameter hole in the forms, and insert a section of the pipe or conduit before pouring concrete.

While formwork requires the skills of carpentry, it is not necessary to build forms that are neat and trimmed. Forms

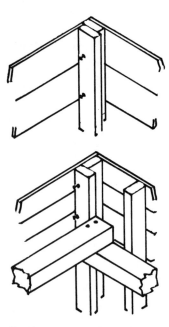

Figure 62. Forms built with double-headed nails to facilitate removal

and reinforcement may well cost more in material and labor than the concrete that makes up the wall. Consequently, lengths of lumber used for wales may be random with staggered splices. If they extend beyond the form, it is not necessary to trim them, only to nail them and brace them securely.

Remember that the forms will be removed, though. If plywood corners overlap, one of the sheets will wind up embedded in the mix and be difficult to strip. Don't worry about the outside form surface. It is the one you see, but not the one that determines the shape of the concrete.

Foundation Construction and Block Fundamentals

Also, to facilitate removal, the inside form surface should be oiled. Many proprietary formulas are available (usually called form release agents), although the basic ingredient should be refined, pale, paraffin-base mineral oil.

PREFABRICATED FORMS Compared to all the work of a site-built form, though, economic considerations may dictate the use of prefabricated forms (Figure 63). Many types are available.

Unframed $1\frac{1}{8}$-inch plywood panels may be used with a modular system of locking clips panel-to-panel, ties, and metal bracing. Steel channel frames, 3 to 5 inches thick may be used with $\frac{1}{4}$-inch steel plate, or $\frac{1}{2}$- to $1\frac{1}{4}$-inch plywood. Metal-faced forms, particularly aluminum, release easily, although a coating of machine engine oil should be applied prior to concrete placement.

From the contractor's point of view,

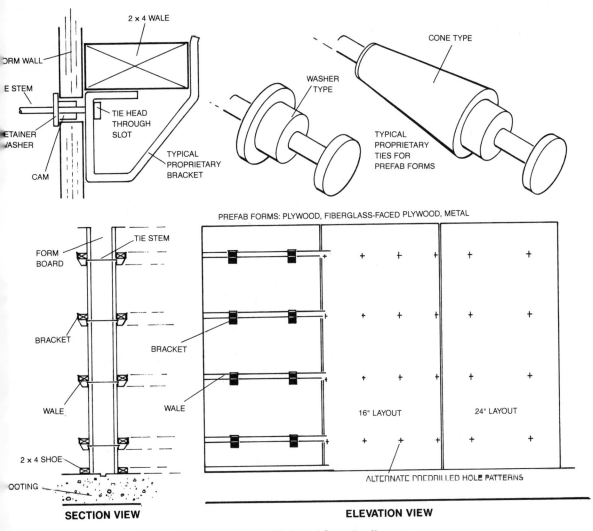

Figure 63. Prefabricated form details

metal forms are ideal for reuse, although they require prompt cleaning to prohibit rusting. Aluminum forms, which have the extra advantage of lightness, may be embossed to transfer brick, adobe, or other patterns and textures to the concrete surface. Prefabricated forms are used primarily by contractors, but if you want to follow up on this idea, contact a large commercial masonry equipment supply house.

Concrete for Foundation Walls

Concrete in foundation walls must be prepared and placed with four primary considerations in mind: to minimize cracking caused by shrinkage in drying, to avoid segregation of the mix, to promote uniform texture and strength with mix consolidation, and to achieve ultimate concrete strength and durability with complete curing.

In residential applications, the concrete should have a design compressive strength of 3,000 psi, and 3,500 psi if exposed to severe weather. A conventional mix would be proportioned as follows: five bags of cement per cubic yard minimally (increased to a six-bag mix for higher strength or where reinforcement is used), with an additional $\frac{1}{4}$ sack for 1-inch maximum aggregate, and an additional $\frac{1}{2}$ sack for $\frac{3}{4}$-inch maximum aggregate (see page 53). Protection from freeze-thaw cycles is obtained by air entraining (typically 5 percent plus or minus 1.5 percent).

In cold weather (below 40°F), a maximum of 2 percent calcium chloride accelerator may be useful to achieve early strength gain. More basic precautions should include the removal of all ice or snow from the footing surface and the form walls.

Although steel reinforcement is generally not required except in extraordinary circumstances (designing in earthquake zones, for instance), it can minimize cracks that form due to shrinkage in drying. There are several ways to control this cracking: first, by using a high strength mix; second, by using control joints at 15-foot intervals and not more than 10 feet from corners (joint depth roughly $\frac{1}{5}$ of wall thickness—Figures 64 and 65); third, by running two Number 4 rebars at the top of the wall and immediately below window openings; and fourth, by adding rebars according to the following table.

Location of Number 4 Rebars for Crack-Resistant Walls

Wall Thickness (inches)	Number of Bars (top of wall)	Bar Spacing and Location (inches)	
		Space between Bars	Distance from Form
6	1	18	2
8	2	14	3
10	3	11	4

NOTE: top bars should be placed 4 inches below the top of the wall.

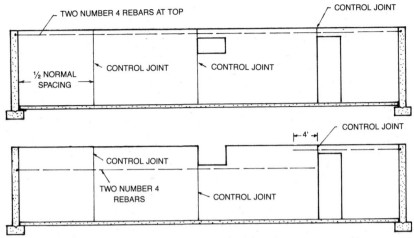

Figure 64. Control joint and rebar locations in perimeter walls

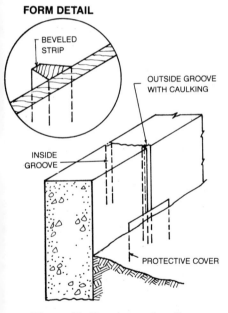

Figure 65. Forming and sealing control joints in poured walls

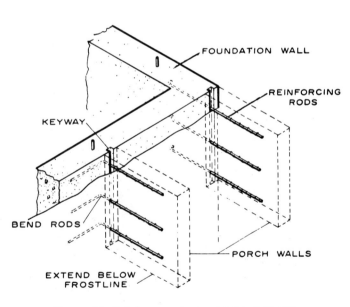

Figure 66. Reinforcement at entrance step supports to prevent settling

Segregation of the mix (the separation of ingredients), another factor that can weaken concrete and promote cracking, can be avoided by placing the mix immediately over the forms. When concrete migrates in the form more than 2 or 3 feet, or when it is dumped against the form wall instead of directly down on the footing, segregation occurs.

Consolidation, which assures a smooth surface and a uniform mix, can be accomplished by hand but is easier and more effective by vibration (Figure 68). An emersion-type vibrator, held nearly vertical for about ten seconds at each point, produces a surface free of honeycomb and air holes. The vibrator should be moved when entrapped air bubbles no longer break the concrete surface, or when the surface becomes relatively flat and glistening. At this point the vibrator motor will return to normal speed after the initial strain of entering the concrete.

While vibration produces excellent concrete it also alters the affected area so that pressure against the form walls approaches the full pressure of a liquid weighing 150 pounds per cubic foot. Typical residential formwork should be able to withstand these loads, but in combination with a full placement of concrete in the forms, it may prove to be too much.

Lateral pressures can be controlled by placing the mix at a rate up to 4 or 5 vertical feet per hour in separate layers, called lifts. Each lift should be made in 12- to 18-inch layers, without a significant time lapse that could produce cold joints between layers. Air temperature

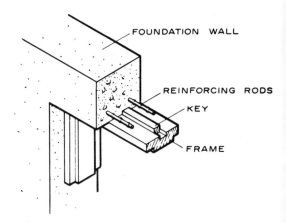

Figure 67. Reinforcement over openings to prevent cracking

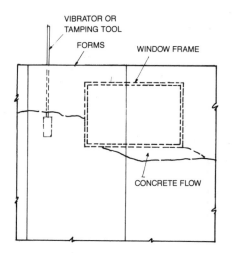

Figure 68. Consolidating the mix under a formed opening

should be considered as well, because it affects the lateral pressures during placement. For example, at a placement rate of 4 vertical feet per hour, a lateral pressure of 550 pounds per square foot is exerted in 90°F temperatures, while at 40°F the pressure is nearly doubled.

Temperature will also affect the curing rate of concrete foundation walls. If the forms are exposed to direct sunlight they must be kept moist (see page 59). The forms should be left in place at least two days in moderate weather and a week in below freezing temperatures.

After the forms are removed, concrete should be cured until it is at least seven days old by soaking or covering with curing compound or plastic sheeting (see page 60).

WOOD FOUNDATION ALTERNATIVES

The residential construction industry is always searching for new building materials, and new ways to combine existing materials and technology into efficient building systems. Wood foundations (notated AWWF, for "all-weather wood foundation"), are a case in point. The materials are commonplace, consisting of construction quality framing members and plywood, with one important addition. All members placed below grade are pressure-treated with wood-preserving chemicals.

The motivations for using an AWWF system are, first, to reduce foundation costs, and second, to speed construction. Surveys of several residential building firms using AWWF have shown a reduction in costs of about $400, compared to conventional masonry foundations, on an average, 1,500-square-foot development house. As to construction speed, foundation panels, fabricated in the shop using pressure-treated lumber and stored at the site, have been erected in less than three hours, including a panelized first floor and basement stairway.

With AWWF construction, conventional excavation procedures are followed until the basement or crawl space floor level is reached. Then a sump pump and all plumbing lines are laid in a bed of gravel or crushed stone. Beneath the perimeter walls, the stone bed must reach below the frost line; but the footing is simply a 2 × 10 or 2 × 12 pressure-treated timber, laid directly on the stone. The panelized walls rest on this sill, and when plumbed, squared, and securely braced, may be backfilled.

Typical AWWF construction, from bottom to top, would include the following: 1) a 20-inch-wide gravel trench extending below the frost line; 2) a 2 × 10 footing plate; 3) a 2 × 6 stud-wall shoe; 4) a 2 × 6 stud wall, 16 inches on center, filled with R-19 or greater fiberglass insulation; 5) a 2 × 6 top plate; 6) a facing of ½-inch C-D, exterior grade plywood; and 7) a 4-mil or thicker polyethylene covering pinned just above the finished grade with a trim board.

This basic structure may be modified by adding a knee wall (a short wall built against the foundation) to support brick veneer, and, of course, height is adjustable to accommodate a crawl space or full cellar. The panels are locked by adding a second top plate (not made of treated lumber), and continuing with conventional first floor framing.

Although burial tests for treated plywood over extensive periods (during which preservative technology has been improved) have shown little or no deterioration, the AWWF system must still be considered a semiexperimental technique, and even then is limited to commercial operations that have shop facilities for efficient panel assembly.

The system has proved reliable in many diverse site conditions, but before using it examine the soil type and drainage characteristics of your site carefully.

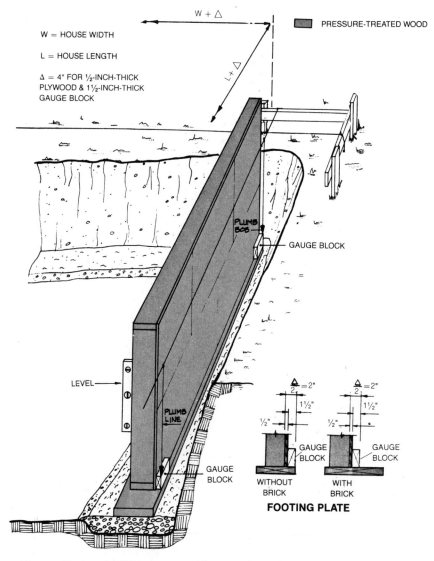

Figure 69. An AWWF system with treated wood footings on a gravel bed

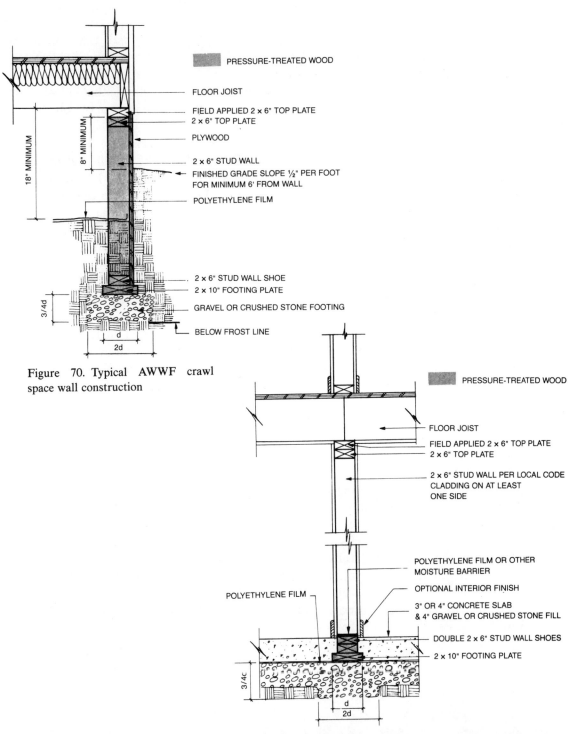

Figure 70. Typical AWWF crawl space wall construction

Figure 71. Typical AWWF center bearing wall in basement

WEATHERIZING FOUNDATION WALLS

Probably the most common complaint about houses, new or old, is dampness or water in the basement. After all the care taken with concrete mixes or mortar joints, somehow, after a few seasons or a few years, water finds its way through the "impenetrable" foundation wall.

Impossible but true problems like these are characteristic of building, and the best motivation for thoroughness and accuracy in each step of construction. Once the foundation is built, the house resting on it completed, and the soil around it backfilled and landscaped, leaks in the foundation wall can be difficult to pinpoint and expensive to correct.

The solution that works most effectively at this point is to dig trenches along the house to expose the wall (backfilling in reverse), and to install perimeter drain tile at the footing to foundation seam, or to apply a waterproof coating on the wall face, or a vapor barrier, or to spend the time and money on all three weatherizing systems.

Obviously, it makes more sense to build in these extra measures of protection (even though the concrete or block should be sufficient on relatively dry sites) while the foundation walls are exposed. There are four basic ways in which foundations should be protected: 1) with drain tile to carry groundwater away from the house; 2) with waterproofing or vapor barrier surface applications to keep water and moisture out of the masonry wall; 3) with insulation to reduce heat loss and radical temperature variations on each side of the wall that can foster condensation problems; and 4) with termite shields to prevent these destructive insects from reaching that ideal food source—the wooden frame of your house.

Drain Tile

Drains should be installed at or below the area to be protected. Prime locations are on the footing next to the foundation wall, next to the footing, or, on sites where the water table approaches floor slab level, inside the footing just below the finished floor (Figure 72).

Clay or concrete drain tile (4-inch diameter and 12 inches long), should be placed on a 2- to 4-inch gravel bed with a slope of at least $\frac{1}{2}$ inch in 12 feet, to carry groundwater to a storm drain (consult local codes), natural runoff, or a dry well.

Tiles are placed $\frac{1}{8}$ to $\frac{1}{4}$ inch apart, and the joints are covered with flaps of asphalt-saturated roof felt, secured with tie wire (Figure 73). Water will seep into the drain through these joints and run off. A layer of gravel 6 to 8 inches deep covers the tile, followed by compacted fill. Continuous lengths of perforated plastic or Orangeburg pipe may be used as well, with perforations down.

Waterproofing and Vapor Barriers

First, use a patching mortar (one part cement, two parts sand) to fill any voids in the foundation, for example, holes left after tie removal on poured walls (Figure 74). After patches have cured, select one

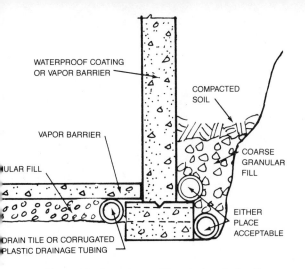

Figure 72. Locations of foundation drain tile

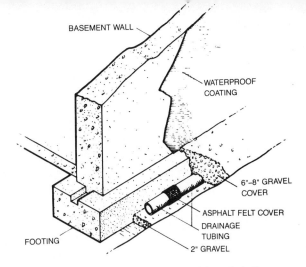

Figure 73. Foundation drain tile with asphalt flaps over joints

Figure 74. Patching mortar is used on cracked or open joints.

of several membrane materials to seal the foundation wall. In most cases membranes are malleable enough to cover the joint between footing and foundation, carrying water directly to drain tile laid even with the footing base.

The membrane acts as an envelope around slab and foundation to seal out water and moisture that can cause bad odors, paint peeling, mildew, rusting, loose floor tile, and other problems. There are several types. Roofing felt, with 6-inch sealed overlaps, and polyethylene film (easily punctured during construction), may not be permanent because they can be disrupted during simple operations like opening a crawl space vent or fixing a plumbing leak.

After curing, good results can be obtained with prefabricated asphalt membranes (frequently laid in combination with an impaction strip between footing and foundation), asphalt-plastic membranes, butyl-rubber sheeting (more expensive), and similar materials, bonded to the concrete with rubber-asphalt adhesive and using 6-inch wide gusset strips of the same material over vertical seams. Hot-mopped asphalt can be used as well, alone, or as a final seal over a $\frac{1}{8}$-inch thick asphalt membrane, generally available in rolls.

The effectiveness of these membranes is

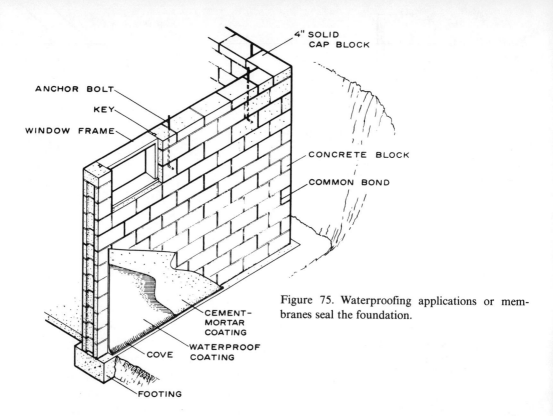

Figure 75. Waterproofing applications or membranes seal the foundation.

determined by the completeness of the envelope (using premolded membrane collars around soil pipes, for example), and by the perm rating of the material, (its permeability to moisture—the lower the rating, the better).

Foundation Vapor Barriers	
Membrane Type	*Perm Rating*
15-pound roofing felt	0.6–2.0
4- and 6-mil polyethylene	0.10
55-pound roofing felt	0.03–0.08
Rubberized, asphalt-coated polyethylene	0.03
Butyl rubber sheeting	0.002
⅛-inch asphalt panels (or asphalt-plastic)	0.00

Insulation

It is not uncommon for homeowners with wet basements to mistake severe condensation problems for outright leaks. If the air inside the cellar is warm and carries moisture—all interior air does, from washing, cooking, and other household operations—the moisture may condense on the cool foundation wall. For example, air 70°F and 40 percent relative humidity has a dew point (the point of condensation) of 45°F.

The solution, in addition to directly venting as much moisture as possible from kitchens and baths, is to keep the temperature at the inside surface of the foundation wall above the dew point by designing the walls with a low heat trans-

Dew Point for Given Temperature and Relative Humidity

Room Temp. (degrees F)	Dew Point at Relative Humidity Percentage (degrees Fahrenheit)									
	10%	20%	30%	40%	50%	60%	70%	80%	90%	100%
50	−1	13	21	27	32	37	41	44	47	50
55	3	17	25	31	37	41	45	49	52	55
60	6	20	29	36	41	46	50	54	57	60
65	10	24	33	40	46	51	55	58	62	65
70	13	28	37	45	51	56	60	63	67	70
75	17	31	42	49	55	60	65	68	72	75
80	20	36	46	54	60	65	69	73	77	80
85	23	40	50	58	65	70	74	78	82	85
90	27	44	55	62	69	74	79	82	86	90

NOTE: convert Fahrenheit to Celsius by subtracting 32 and multiplying by 5/9.

mission rate, called a U-factor. While R-values rate the thermal value of insulation, the U-factor accounts for the thermal properties of every wall component (even dead air spaces), for an overall thermal performance rating.

The table and graph on this page are used to find the dew point and U-factor required for a given room air temperature and relative humidity.

1) On the table above, find the dew point for room air temperature and relative humidity. Example: at 70°F and 40 percent relative humidity, dew point is 45°F.

2) Find the difference between the inside and outside temperatures. Example: 70°F inside, −10°F outside, difference of 80°F.

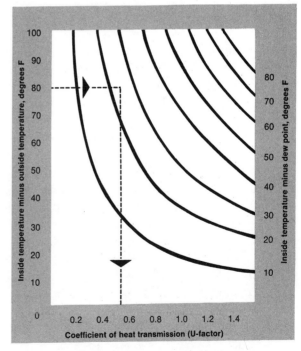

Figure 76. U-factor is calculated using inside and outside temperatures and dew point.

U-factor Computation

Without insulation		With insulation	
Wall Component	R-value	Wall Component	R-value
inside air film	.68	inside air film	.68
8" concrete block	2.00	½" gypsum wallboard	.55
¾" air between furring	.91	1" styrene board	6.20
exterior wood siding	1.00	8" concrete block	2.00
		¾" air between furring	.91
		exterior wood siding	1.00
	4.59		11.34

$$\text{U-factor} = \frac{1}{R+R+R\ldots} = \frac{1}{4.59} = 0.22$$

(without insulation, does not come down to FHA .07 standard)

$$\text{U-factor} = \frac{1}{R+R+R\ldots} = \frac{1}{11.34} = 0.8$$

(with insulation, reduces close to FHA .07 standard)

NOTE: the higher the total R-value, the lower the U-factor.

---

3) Find the difference between the inside air temperature and the dew point. Example: 70°F inside, 45°F dew point, or a 25°F difference.

4) On the graph, find the intersection of the inside/outside difference (80), and the inside/dew point difference (25).

5) Read the U-factor required to prevent condensation directly below this intersection. Example: 80°F from the left column intersects 25°F curve from the right column over .52 U-factor.

Bear in mind that the U-factor needed to prevent condensation is a minimal number intended to prevent deterioration and minimize maintenance. For high energy efficiency the U-factor should be reduced below this level. To achieve efficient U-factors in your climatic zone, consult your building department or local U.S. Department of Housing and Urban Development field office for current FHA minimum standards of U-factors in new homes. In an area roughly between northern New York State and Virginia, FHA U-factor standards are .03 in ceilings and .07 in walls and floors. This factor is computed as the reciprocal of all R-values in the wall, i.e.,

$$U = \frac{1}{R + R + R \ldots}$$

A plain, 8-inch-thick concrete wall has a U-factor of 0.70 ($8 \times 8 \times 16$-inch concrete block with voids has an R-value of nearly 2.0), but this can be reduced dramatically by installing a 1-inch-thick

Foundation Construction and Block Fundamentals

tongue-and-groove, rigid insulation either inside the wall or outside it, covered by the membrane (Figure 77). The following example shows how to compute U-factors for a block foundation.

Never buy insulation based on thickness; always evaluate it by the R-value, which must be stamped by regulation on all fiberglass rolls and bats and all bags of loose fill insulation. Fiberglass insulating board has an R-value of 1.8 to 2.2 per inch. Rigid foam has an R-value of 3.7 per inch for polystyrene and up to 6.2 per inch for some expanded polystyrene and polyurethane boards. Note that most codes require interior foam board to be covered with wallboard for fire retention.

Termite Protection

Although obviously not a form of weatherizing, termite protection is essential in most areas. There are two types of termites that cause destruction: dry-wood and subterranean. Dry-wood termites inhabit only the most southern, coastal sections of the country. Intensive application of preservatives, dense screen coverings, and the use of special grades like heart

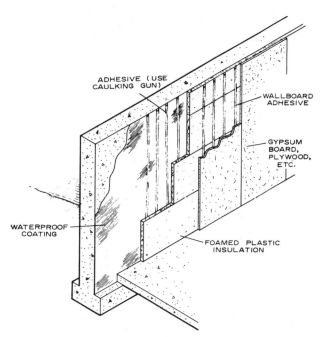

Figure 77. Sealing and insulating the foundation interior skin

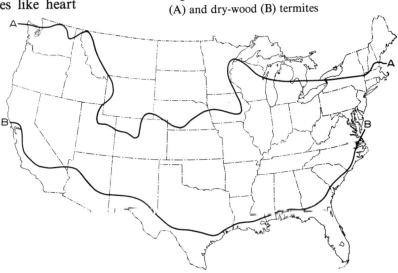

Figure 78. Northern limits of subterranean (A) and dry-wood (B) termites

redwood offer protection against these termites, which fly directly to the food source and bore into the wood.

Subterranean termites, on the other hand, are found in all but the northernmost sections of the country, and either mountainous or arid regions (Figure 78). They thrive in warm, moist conditions, and build mud tunnels (¼ to ½ inch wide), between their shelter and their food source.

Protection includes removing stumps and wood debris from foundation fill, and sealing any cracks in the wall that could serve as entry points. Soil poisoning may be used (emulsions such as 5 percent aldrin or dieldrin used to be common), at a rate of 4 gallons per 10 linear feet of foundation, remembering that some poisons like chlordane, which is potentially lethal to your pets as well as your termites, may not be permitted by local codes or the Environmental Protection Agency, whose Office of Pesticide Programs evaluates and registers every chemical pesticide.

A final line of defense is provided by an aluminum termite shield laid on asphalt adhesive, and extending an inch beyond each side of the foundation wall with edges bent down toward the ground (Figure 79). Pay particular attention to perforations in the shield made by anchor bolts, sealing them with surface asphalt if necessary.

Backfilling

After all foundation work is completed, clean fill should be replaced against the foundation, allowing for a final grade that will slope away from the house. First, remove construction and organic debris from the trench. Then, to minimize settling, place and tamp fill in 2-foot thick layers.

Bear in mind how difficult and costly it is to solve foundation problems after the house is completed and landscaped. Now is the time to provide for structural durability and long-term protection against the elements.

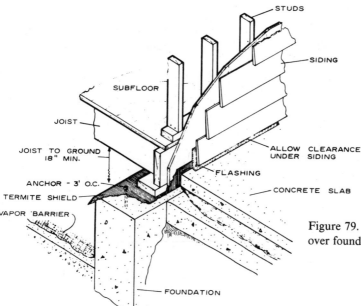

Figure 79. Metal termite shield with extension over foundation wall

6

SLAB CONSTRUCTION

At this stage of construction the site will begin to look a little worn, with materials stacked on skids to one side of the driveway, empty cement bags floating about in the wind, and stakes left over from form-bracing dotting the muddy fill around the foundation. After all this work you still have a hole in the ground or, if it has rained recently, a small lake, boxed in with naked-looking foundation walls. But the bottom of the box is missing.

To complete the isolation of the house from the ground, this missing piece must be fitted in. And just as the bottom of a grocery bag is likely to be its weakest link, most basement and crawl space problems are apt to come through the floor, not the solid, weatherized foundation.

In the majority of houses the footings and foundations get most of the attention. It is easy to perceive them as a wall, almost a battlement, against soil, moisture, water, termites—all the forces that attack a house, undermine it, and cause deterioration. Compared to this heroic effort, a concrete cellar floor or soil in a crawl space seems inconsequential, and may be treated as an afterthought—just filling in the bottom of the box.

To a large extent this is all right for crawl space designs. Preparing the final interior grade is an inexpensive, straightforward operation that permits first-floor framing to proceed immediately. However, it is not so simple for slab construction, whether at grade or on the cellar floor. In these cases, the masonry envelope must function as the foundation walls do with only two differences: They need not carry severe structural loads, and they must seal the house from the soil below it instead of around it.

Design Alternatives

Concrete slabs can be laid just above the footing to serve as the finished floor for a full-height cellar. They can be laid close to the finished grade with thickened edges to serve as footing, foundation, and first floor all rolled into one. They can be eliminated, in crawl space construction, and replaced by a moderate amount of soil preparation and a vapor barrier.

Full cellars offer many obvious advantages but require careful construction and weatherizing. Crawl spaces offer other advantages, including speed of construction and minimal maintenance. Slabs on grade, however, require roughly the same amount of material and labor as basement floors without providing the advantage of an additional, although partially subterranean, story. In fact, slab-on-grade construction is likely to breed more maintenance problems than similarly constructed basement floors simply because soil, moisture, water, insects, and other deteriorating influences are so close to finished, interior space.

As the cost of new homes becomes prohibitive, however, the luxury of a full cellar becomes expendable. It is an easy part of the design to lop off. Without it, costs and price tags are reduced. And this seems to be why slab-on-grade construction is still popular. Even with the potential maintenance headaches, the enticing savings of constructing the footing, foundation, and first floor in a single concrete pour continues to produce slab houses.

Design Requirements

Regardless of economics, sites with soil that is marginally porous, or with poor drainage in general, will not accept slabs on grade without problems. And although construction details and structural requirements for slabs on grade and slabs on the cellar floor are similar, the functions they perform in these locations are not.

Typically, cellar floors are left exposed in raw concrete or painted. If there are problems with condensation, the unfinished floor will not be damaged. But on the first floor, painted concrete is not an acceptable finish. When slab on grade building first became popular, the solution was to apply asbestos or vinyl tile directly to the concrete.

It is surprising that so many people settled for this type of finishing, because the most elementary rules of weather forecasting indicate that when something warm meets something cold, and one of them is holding water vapor, condensation will result. Without insulation to reduce temperature differences and vents to evacuate moisture, the tile would sweat profusely during summer months, when the ground beneath the slab was cool and the air above it was warm and moist. When, on many designs, this problem persisted in the winter, radiant heating (pipes installed in a pattern throughout the slab) was used to prevent condensation by keeping the slab warm straight through to the ground.

This type of construction can still be found on some houses, and, not surprisingly, on motels that were built in the 1950s and early 1960s. Construction time and quality were cut with skin-deep excavation, a layer of gravel, heating pipes,

Slab Construction

and a concrete floor. Unfortunately, though, the more finishing materials you applied over the concrete (carpeting, for instance), the less heat radiated into the room. And when winter temperatures plunged, the concrete floor could become too hot to walk on with bare feet.

This leads to the requirements of a modern slab on grade house. First, and most important, although the correct term is "slab on grade," the principles of foundation grading must still be applied, namely, that topsoil will slope away from the house for at least 2 feet, and that siding, which should extend past the 2 × 4 shoe to cover the seam between frame and foundation, should be at least 8 inches off the final grade. The slab (normally 4 inches of reinforced concrete) may require excavation of loose topsoil and carting in of clean, dense fill, which must be compacted before provision for utilities is made and a 4- to 6-inch gravel or crushed stone bed is laid. In any case, pouring a slab even with or, worse yet, below the finished grade will create a shallow pool in summer and spring, and an ice rink in winter.

In planning a slab on grade, careful calculations must be taken for waste systems, for example, the location of the soil pipe, its slope through the slab, point of exit through the thickened-edge foundation, and this elevation in relation to town sewers or a private septic tank and leaching field.

Other requirements for a relatively maintenance-free slab are 1) a vapor barrier with a low perm rating below the concrete; 2) rigid insulation at least along the perimeter; 3) the use of reinforcement to minimize cracking; and 4) careful floating of the mix to create a smooth, dense surface (see pages 104–108).

CONCRETE MIXES

Concrete for slabs should have a low water-to-cement ratio, and hold as much coarse aggregate as possible at the surface. This type of mix provides high strength and water resistance, and should achieve a compressive strength of at least 3,500 psi. A mix using 1-inch aggregate should contain at least 520 pounds of cement per cubic yard, while a ¾-inch aggregate mix requires at least 540 pounds.

Admixtures that combine water reducing and either retarding or accelerating properties may be used to allow more water in the mix initially for better workability. Remember that air entraining (page 54) is needed for freeze-thaw protection on slabs as well as footings and foundations.

Forming the Pour

Slabs on grade require conventional forms, accurately positioned and sturdily braced. Basement floor slabs are already formed by the foundation walls. If perimeter heating ducts are planned, they must be set firmly to resist the force of the pour. Asbestos-cement ducts can be leveled and secured in a sand bed, while metal ducts should be encased in concrete.

Screed boards should be set around the

foundation walls (1-inch thick material tacked to the mortar joints will suffice), with their top edges level with the finished floor. Furthermore, at all points where the slab touches foundation walls, piers, columns, soil pipes, or other obstructions, an isolation barrier is needed to prevent cracking. For instance, at the point where lally columns contact the slab, compressive loads create stresses different from those on other sections of the slab away from the column. Two different stresses on the same material in the same plane results in cracking. So to prevent it, column and foundation wall loads are transferred to their own footings by isolating the loads they carry from the slab. Metal collars or even a strip of roofing felt can be used to prevent bonding.

VAPOR BARRIERS AND INSULATION

Both of these weatherizing systems must be installed before pouring the slab. Vapor barriers, laid over the gravel or crushed stone bed, must have a low perm rating, be unaffected by a buried environment, and be able to maintain their potential after the pour is made.

A high-quality slab transmits about one quart of water per day per thousand square feet if the soil it rests on is moist. While polyethylene film has a low perm rating, it is easily punctured by aggregate in the concrete or by pressure from the reinforcing wire during construction. Triple layers of 55-pound hot-mopped roofing felt, or other heavy duty membranes (see page 96) offer more durability, and should be continued around the slab edges to join the membrane on foundation walls.

Insulation, at least along the slab perimeter, must meet similar requirements, particularly a resistance to crushing from the concrete load. Many types are available, including rigid fiberglass board and foam board. The table on page 105 indicates recommended R-values (resistance to heat transmission) according to average low winter temperatures.

Together, unbroken vapor barriers and insulation can prevent most deterioration caused by moisture. But keep in mind that interior-generated moisture can also cause serious problems. And when the slab temperature is cooler than the dew point for interior air, condensation will result. The slab may be warmed with radiant heating, baseboard units, or other sources. But a well-insulated slab (6 inches of concrete has only a .78 U-factor) will make the heat source work less. (See pages 96–99 for details of dew point and U-factors.)

REINFORCEMENT AND CONTROL JOINTS

When concrete is poured, it is plastic and water heavy. As the concrete hardens, the water acts on the cement and evaporates. And when materials loaded with water dry out, they generally shrink a little. On a basement-size concrete slab that contains large amounts of water, uncontrolled shrinkage can lead to cracking.

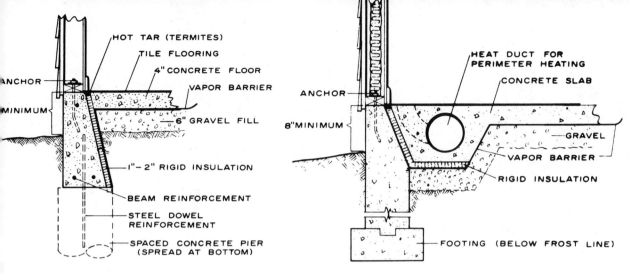

Figure 80. Perimeter insulation

Figure 81. Sealing and insulating the cellar floor slab

	Slab Insulation R-Values		
Average Low Temp. (degrees F)	Insulation Depth Down Foundation (feet)	R-Value No Floor Heating	Floor Heating
−20	2	3.0	4.0
−10	1½	2.5	3.5
0	1	2.0	3.0
10	1	2.0	3.0
20	1	2.0	3.0

While low water-to-cement ratios will help, control joints or reinforcement or both will help keep the slab in one piece.

Typically, gravel is raked roughly level, screed boards are placed, and reinforcing wire is unrolled and walked on to flatten it out before the concrete truck arrives. You do get reinforced concrete this way, but it won't last very long.

First, remember the distinction between control joints cut in the concrete surface to control cracking and isolation joints. Around the slab perimeter, space occupied by the vapor barrier and insulation should not be filled with concrete. This would bind the slab to the foundation wall. Instead, this gap (a built-in isolation barrier) should be filled with a cellular rubber or polyethylene backer rod, covered with elastomeric sealant.

Control joints, on the other hand, should be spaced approximately 15 to 20 feet apart, dividing the floor area into squares, if possible, rather than rectangles. These joints, which can be cut easily with a hand jointer guided against a

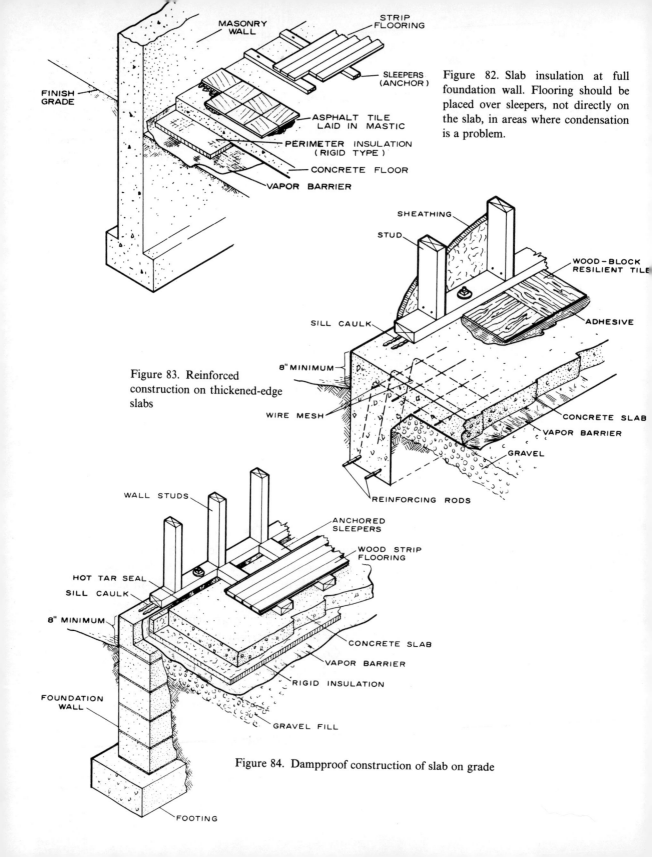

Figure 82. Slab insulation at full foundation wall. Flooring should be placed over sleepers, not directly on the slab, in areas where condensation is a problem.

Figure 83. Reinforced construction on thickened-edge slabs

Figure 84. Dampproof construction of slab on grade

Slab Construction

straightedge, should be cut to a depth one-fifth as great as the thickness of the concrete. Superficial joints are ineffective.

Frequently, reinforcing wire is used without the extra protection of control joints, because the joints give concrete an outdoor, utilitarian look, and because their depressions cause problems (particularly with first-floor slabs on grade) as tile is laid. Control joints may be omitted as structurally unnecessary, but they should not be passed over for these reasons. Control joints will not remain effective if they are breached by reinforcing wire; however, they will retain effectiveness if leveled with a cellular rope covered by elastomeric caulking. Furthermore, in some areas of the country where condensation can be a problem (particularly the Northeast), it is a good practice to cover the slab with sleepers (2 × 4's laid on the flat over a membrane) and subflooring before installing tile, carpet, or other finished flooring.

Reinforcement for slabs may be provided with rebars, but can be accomplished more economically with reinforcing wire, technically, welded-wire fabric, available in rolls. FHA minimum standards call for 6 × 6–10/10 wire on slabs up to 45 feet, 6 × 6–8/8 up to 60 feet, and 6 × 6–6/6 on slabs up to 75 feet.

Welded wire is very effective if it is properly placed. Frequently it is not. If control joints are used the wire must be cut back 2 to 4 inches on each side of the

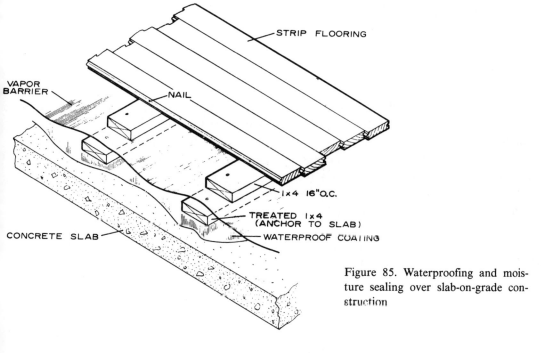

Figure 85. Waterproofing and moisture sealing over slab-on-grade construction

joint and from the edge of the slab as well. The wire should be secured at the horizontal midline of the slab. It is nearly useless running at the bottom of the slab. Since the material is sold in rolls, it should be laid out on the gravel bed and flattened with block or other materials to get the bend out. The wire may be supported on molded concrete units, called chairs, made expressly for this purpose or by pouring the slab to half its final depth, rough screeding, laying on the flattened wire, and completing the pour.

SLAB FINISHING

After the concrete is poured and screeded to a level surface, several finishing operations must be accomplished before bleed water accumulates on the surface. If necessary, coarse aggregate should be tamped or rolled to depress it into the mix surface. Then the surface should be worked with a bull float or a darby (bull floats cover more ground but may not produce as smooth a surface), to reduce high spots and fill shallows. Preliminary joint cutting and edging may be done at this stage.

When bleed water has risen and evaporated, final finishing is started. Early finishing works excess water into the sand at the mix surface, creating a weak, soft skin on the mix. As a guide, wait until the surface sheen of water has evaporated, and the concrete can withstand foot pressure leaving no more than a $\frac{1}{4}$-inch indentation.

The first process, floating, should begin at the slab edges since they harden before the interior, to set aggregate barely below the surface and to consolidate the mix to a dense skin. The convention is to use wood floats on lean-mix concrete without air-entraining agents, and either aluminum or magnesium floats on rich, air-entrained mixes that might stick to wood floats.

The second finishing process, troweling, increases the consolidation of fine aggregate (and the surface strength), while providing a finished, smooth surface. Finally, jointing and edging, if desired in addition to wire reinforcement, should be completed while troweling.

There is no absolute rule used to determine waiting time. This is one point in construction where careful attention and presence at the site are required to get the most out of the building materials. Some air-entrained mixes will not bleed much surface water, and others may bleed so much in conditions that retard evaporation that puddles must be reduced by pulling the water off the slab with loops of hose or by sponging it up with burlap. However, the surface should never be dried by adding dry mix.

Curing

When the temperature is between 50° and 70°F, curing should be continuous for seven days. At higher temperatures five days should be sufficient, although in extremely hot weather, retarders may be needed to slow evaporation and permit thorough hydration and strength gain.

Do not be tempted to rush ahead with construction at the expense of complete curing. Without it, concrete loses strength equivalent to the reduction of cement in the mix by 120 pounds per cubic yard.

Of the three common curing methods, wetting burlap coverings requires the most attention and runs the greatest risk of uneven treatment. Securely lapped polyethylene sheeting is economical, but can produce a mottled discoloration of the surface due to condensation forming under the film. For highly uniform curing, liquid membranes may be applied in two coats with a roller or spray gun (apply the second coat at right angles to the first).

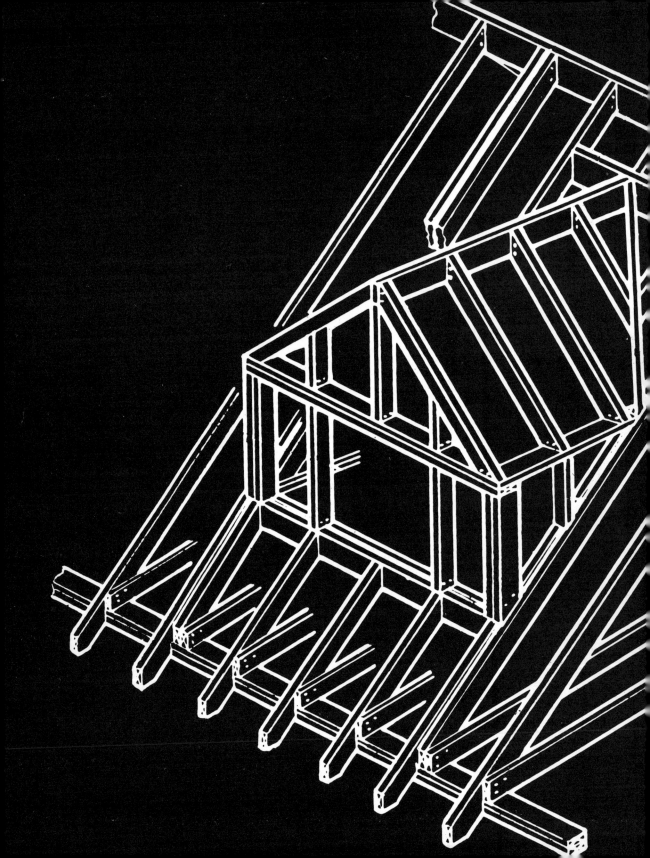

PART 2
FRAMING

7

THE NATURE OF FRAMING

The natural structural systems of the universe and of man are curvilinear. Examples are all around us. The action of the planets is elliptical. The atmosphere is a circular envelope in constant three-dimensional change. The flattest-looking stretch of land bends with the curvature of the earth. The trunks of trees, though they may appear straight from a distance, bow in and out. Even individual, microscopic wood cells that appear rectangular are, in fact, somewhat egg-shaped on the order of cells in a beehive.

Yet with endless models of curvilinear systems and structures in nature, the first tenet of framing construction is the straight line. This artificial concept, which exists only for man's convenience, is the determining factor of residential framing and design. This fact is worth some thought because the straight line influences every aspect of construction from aesthetics to cost. And while, on the surface, the simplicity of the straight line seems to be a convenient design and construction tool, it establishes a fundamental conflict between nature, which is curvilinear, and man-made shelter, which is linear.

CONFLICTS OF DESIGN AND TECHNOLOGY

Often you will hear that a well-designed house fits into its surroundings or blends into the countryside. But does it really? There is no natural form sympathetic to a high ranch, no mirror of repetitive aluminum siding in jagged cliffs. The few houses we judge as complementing nature do so only by comparison to the majority of houses that occupy space regardless, or in spite, of their surroundings. You can see this aesthetic conflict between man-made and natural structures in every part of the country, and particularly in tract developments, whose builders are notorious for turning apple orchards, rock-lined brooks, and rolling grasslands into huge dirt parking lots on which houses are deposited like cookies on a baking pan.

There is simply no naturally occurring facsimile of a typical frame residence. So where do all the straight lines and right angles of residential construction come from? They are the direct by-product of the materials used to build houses. Brick, cinder block, dimensional lumber, and plywood are all flat-sided, square-edged rectangles. These and almost all building materials are linear. Those that aren't, concrete for instance, are hard to use in curvilinear designs because they depend on rigid forms to hold their temporarily plastic consistency. And form materials are linear.

It is, of course, possible to build curves into houses using linear materials, but it is difficult, time consuming, and expensive. This makes curvilinear construction a hybrid design option reserved only for high-priced, architect-designed homes. In essence, building curvilinear shapes with linear materials is a process in which the means work against, not toward, the ends. As a result, the easiest solution is to stand back and let the linear character of building materials dictate the design of the structure.

This doesn't say much for the imagination and initiative of housing designers but, like housing consumers and do-it-yourselfers, they are usually locked in to the materials at hand. But our proclivity for linear buildings is more than a matter of convenience. When you backtrack further along the idea of linear design stemming from linear materials, you will come to the technology used to manufacture and process these materials.

This technology is linear as well. The apparent rationale is purely economic. Curvilinear manufacturing requires handling units of material individually. For instance, seamless, round-edged, fiberglass bathroom fixtures like tubs and showers are made on molds—one mold for one unit. Their shapes are built up by hand, gradually, and at a substantial labor cost. By comparison, linear manufacturing permits the continuous, assembly line handling required to mass produce large-scale materials at a profit. Wood cannot be economically processed one board at a time. To be produced in volume, forests must be cut, and logs must be floated to the mill en masse where they are continuously fed into debarkers and on to the saw. The financial bottom line dictates that wood will be cut into straight lengths, that aluminum flashing will be extruded into long strips, and that plywood, wallboard, paneling, and glass will be manufactured in flat sheets.

Linear technology that produces linear materials will almost inevitably produce linear designs. But this backtracking raises a fundamental question. Is the manufacturing technology linear because that is the best way to produce materials, or is it linear because we find it nearly impossible to do it any other way? This thought seems surprising since all natural models we might copy in construction are curvilinear. Yet our designs are not, our building materials are not, and our manufacturing technology is not. Why? Because we are accustomed to linear relationships, and find it difficult to re-create the incredible efficiency of natural curvilinear systems.

As a very elemental demonstration of this gap in our conceptual ability, pause a moment as you read these words, and think of this book remaining motionless while you move back away from it. This may be easier to accomplish if you close your eyes. Now the book appears at a distance and is smaller. Then reverse the process until the book appears in its true size.

Now attempt to think of yourself slowly somersaulting, head first, over the top of the motionless book—over it, down toward the floor, then underneath the book (without mentally twisting to one side or the other), and back up to your starting position, completing the circle. You can try this conception eyes open or eyes closed.

The first exercise is linear, moving backward and forward in a straight line. It is relatively easy to visualize this movement through your own eyes, or in the first person. The second exercise is curvilinear, and, for most people, it is impossible to visualize in the first person, or through their own eyes. In this curvilinear exercise you can visualize going over the top of the book, but hit a blind spot (really a break in the conceptual somersault) somewhere near the bottom of the circle. It is much easier to imagine this motion from the side, in the second person, as though you were in a theater watching it on a two-dimensional screen.

There are endless examples of our bias to linear rather than curvilinear understanding. We spend years learning a linear number system, but few of us are exposed to theories of relativity, and the mathematics of curved time and space. We read in straight lines, and figure distances in straight lines. We even visualize the state we live in as a flat outline on a flat plane—we picture it as a map—which is an odd way to synthesize the endless number of curvilinear (topographical) images we have of the land around us.

In many cases we seem to have an almost instinctive aversion to curvilinear understanding. If you were asked to demonstrate the structural principles of a table using only everyday objects found around a house, it would not be much of a challenge. You could simulate this linear structural system using three books, children's blocks, sticks of butter, you name it. But consider your level of self-confidence about demonstrating the structural principles of soap bubbles, whose common joints always take the most perfectly efficient, three-dimensional line between bubbles.

We are accomplished linear thinkers, who design linear technology to manufacture linear materials to build linear houses. This deep-seated linear bias severely limits the scope of residential design and locks us into the second important tenet of residential construction—the idea of building a fortress against nature.

THE MAN-MADE FORTRESS

Consider this. When you dig a hole in the ground, and dig deep enough, you will encounter water, or at least moisture. When you build a house with a cellar, considerable time and money and material is used to keep this groundwater out. Then more effort is devoted to digging another hole (your well) so that groundwater can be piped into the house for drinking, cooking, and washing. Then another hole is made into which you drain all the clean, drinkable water that lands on the roof.

Residential construction is laced with these kinds of inconsistencies, redundant components, and systems working at cross-purposes. Why? Because the house is working against powerful and persistent natural forces. To overcome the effects of one element, construction is employed that deals ineffectively with a second element, which is overcome by more construction, ad infinitum. The result is a fortress with complicated, overbuilt, and static defenses against a supple, changing environment.

In recent years this syndrome has been developed into the "tight house" theory. This concept takes the massive Victorian fortress with solid core doors, plaster walls, and 2 × 4-inch studs that actually measure a full 2 × 4 inches, into the energy-conscious, financially strained 1980s. The tight house has a minimal shell and maximum insulation. It decreases window area to cut heat loss, and seals the house so thoroughly that fresh air must be ducted in to support combustion in the furnace and fireplace. Its ultimate conclusion is a completely self-contained, near hermetically sealed environment in which building materials like vapor barriers and weatherstripping become crucial.

In the tight house, widely spaced 2 × 6-inch studs accommodate thick blankets of insulation and other energy saving systems, to form a modern, lighter-weight version of the fortress. Instead of sheer mass it defends against natural forces with complex, energy-efficient materials in configurations that have little margin for error.

Fortress construction, linear materials, and straight-line design are the rules in residential construction. There are exceptions, but they are not fully appreciated or understood for two reasons: first, many curvilinear construction methods rely on the availability of building stone and cheap labor and are, therefore, a thing of the past; and second, many current exceptions to typical linear construction exist only in underdeveloped areas that seem too culturally and technologically remote to be instructive.

Most notable of these so called primitive examples are the many varieties of huts that have straw or leaves woven over bent trees to form dome-shaped roofs, and the most simple home in design and use of material, the igloo. These exceptions have several characteristics in common: They are both built with natural materials and without the aid of linear technology; their shapes act in harmony with natural forces; and their designs are curvilinear. The most obvious characterization, that they are somehow so simple and primitive that they have nothing to show us, would be erroneous in light of the great efficiency of their designs. The absence of technology, the natural harmony, and the curvilinear designs of these exceptions are clear contrasts to the linear structures in our complex, mechanized, and technology concentrated civilization.

Of course cultural influences are significant. Homes must be constructed simply when living patterns dictate that they will be left behind on a regular basis. But why should these simple homes be so much more efficient than our linear buildings? This contradiction is most striking in the igloo. Its individual blocks might just as easily be cut into square-edged, bricklike rectangles. But the slightly curved blocks, shaped into a shallow dome, hold precious body heat, and shed high winds that would blow down a vertical wall of the same material.

Reconsider the distinction between linear technology as the best way to produce building materials and as a limitation of our linear minds. If the arctic regions underwent a boom and became heavily populated, would we be capable of mass producing curvilinear building blocks and other materials for modest, igloo-shaped homes and expansive, igloo-shaped public buildings? Or would we (as we have for many arctic settlements) import our linear building materials and impose their inefficient, static design on an incompatible environment?

The second group of exceptions to linear construction are relatively old masonry structures, the soundness of which often relied on the arch. In fact, brick arches that date to 3000 B.C. have been unearthed in Egypt and Mesopotamia. The arch is a vehicle for turning the design strength, particularly compression or downward force resistance, of a vertical masonry wall into a horizontal structural system.

The idea is disarmingly simple, needs only minor adjustments from vertical to lateral to make it work, and provides an unexpected margin of construction safety and durability. In other words, arches are a little harder to build than vertical walls, but a lot harder to knock down. But these advantages are outweighed by considerations of material and labor. Stonework is expensive.

While there are many ancient or primitive exceptions to straight-line and fortress construction, in modern residential construction there is only one of note. It is the geodesic dome and it is a rarity, still perceived in the public mind (and in the minds of code writers and building inspectors) as an alternate energy, alternate life-style experiment.

In principle it is marvelously efficient, enclosing a large space with relatively little material. It hugs almost any site, and presents a continuous, three-dimensional plane to the forces of nature. Yet as the forerunner of curvilinear design it suffers as a round form executed with the linear materials of concepts it tries to leave behind.

The resulting weak link in dome construction is the roof skin. Typically, it is made of plywood panels (a linear holdover), which break the curvilinear plane of the dome into sections like facets cut on a diamond. These flat sections form joints that are very likely to leak and deteriorate. This problem pinpoints the contradictions of using linear materials in curvilinear designs.

It also highlights a major deficiency in modern building materials. While almost all natural examples of exterior skin are resilient, even under extremes of temperature and moisture, there is no flexible, resilient, weathertight material for house exteriors. This lack of a curvilinear building skin helps to perpetuate fortress and straight-line construction. Why build structurally efficient, curvilinear, or worse yet, flexible frames if there is no complementary material with which to cover them? Why design a house that is supple against the forces of nature if flashing, chimneys, skylights, windows, doors, and other components are rigidly linear?

BREAKING DESIGN BARRIERS

These questions demonstrate the self-limiting character of residential construction today. Some design and material boundaries are being broken, but the progress is unsteady and fragmented. Houses, after all, are the repositories of many complex technical, political, economic, and sociological issues. A house with massive concrete, direct heat gain walls, and elaborate solar capacity may be on the technological forefront of residential energy use. It may also be economically unfeasible. A house that is carefully and individually planned for family use, with provisions for privacy and communal activity, with spaces for indoor and outdoor recreation, and with areas that provide an environment for individual cultural patterns, may be an energy sieve, or beyond the capacity of small parcel zoning.

The collective quantum leap combining new designs, materials, energy use, and other factors has not been made. And as priorities change, and new issues emerge from decade to decade, a definitive, all-encompassing housing solution can never be achieved. But it is a theoretical, if illusory, goal that, as we strive to attain it, helps to improve the quality of time spent in the places where we live.

In spite of the many problems with modern residential construction, we are approaching a point where some fundamental changes will be possible. And as usual, the most potent stimulus for this change is economic pressure. It is increasing exponentially, applying the combined forces of growing energy costs, land values, tax rates, and mortgage rates, along with the costs of conventional building materials, the labor needed to assemble them, and the efforts required to maintain them.

Currently, and in the next decade, economic pressure will produce two kinds of reactions in the residential housing industry. The first is geared to continuing the construction status quo. Mortgages of thirty years and longer, variable rates, and other forms of flexible financing, the no-frills house with smaller rooms, fewer doors and windows, and no appliances included, are all temporary and somewhat short-sighted stimulants whose purpose is to preserve residential construction as we know it.

The second and more realistic reaction to the pressure of housing economics is the development of new building materials like aluminum and vinyl siding, pressure-treated, all-weather wood foundation systems, and rigid foam panels that double as insulation and sheathing over specially constructed frames. Some innovations attempt to increase durability or decrease maintenance requirements. Others combine building operations to save on labor and material costs.

While these construction advances are all products of linear technology, and, therefore, inevitably bound to fall short of a long-range, holistic change in the places where we live, they do foreshadow more fundamental changes of the future. There will be new materials that are compatible with new designs, but these changes are extremely difficult to make in the uncertain atmosphere of economic hard times. How many builders can go out on an innovative design limb when razor-sharp inflation and 17 percent mortgages threaten to cut down the whole tree?

In the next decade, the housing industry will concentrate on making the typical, single-family house more affordable, not necessarily more functional. This decade also offers the realistic possibility of economically efficient solar energy production. And this kind of advance could well be the holistic catalyst, the technological glue, that binds fragmented material, economic, political, and sociological changes together into a new kind of house.

Today, solar energy is converted to practical applications at well under 10 percent efficiency. And the standard conversion tool is the flat plate collector, a linear design determined as much by the house it sits on as by the function it performs. Some equipment is more innovative than this, for instance, focusing collectors that concentrate the sun's rays with Fresnel lenses, or dish mirrors, or conventional optics, and collectors that use sensors and articulated, motorized mountings to track the sun's path through the sky. Also, promising results have been obtained from new, passively solar, architectural forms like direct-gain foundations, and greenhouse-effect masonry walls.

However, the final potent catalyst is likely to be an inexpensive photovoltaic cell, made of some synthetic silicone substitute, which will efficiently convert sunlight directly to electrical power. When technology and industry become capable of delivering power from the immense, curvilinear energy system surrounding us (and think what that will mean to oil, gas, and nuclear-based utility companies), it will signal an end to the constrictions of straight-line construction, and along with it, the linear limitations on our imaginations.

Will houses of the twenty-first century be domes and dishes that incorporate flexible solar, wind, and thermal converters with undulating, seamless, weathertight skins? Possibly. But this description may turn out to be as much of an anachronism as the flat plate collector. In the meantime, while there are many variations of frame design and construction, it will continue to be difficult and expensive to apply anything but straight-line and fortress concepts. Certainly, we are bound by the materials at hand, but an effort can be made to avoid their most limiting factors.

Remember that a frame, even for a small deck or simple addition, starts as a design that occupies or encloses empty, three-dimensional space. And while residential framing may be constrained by considerations of cost, safety, maintenance, energy efficiency, and other matters, empty space is the ultimate clean slate.

FRAMING TOOLS AND EQUIPMENT

Carpentry is an ancient craft. It has been refined over thousands of years in the construction of pyramids, temples, huts, and cabins. Gradually, the products of this craft have become more complex, more laden with technology, and more diverse. These traits have been mirrored in the tools of the trade.

A quick stroll through any well-stocked home center or hardware store reveals a dazzlingly diverse array of hand and power tools. In a good commercial supply house you will discover another cache of tools like ladders, scaffolding, and other construction equipment. And at some point you will discover tool and equipment mail order catalogues with imported saws and one-of-a-kind chisels displayed in full color.

There are hundreds of different tools and a hundred variations of each type. They range from painted dime-store facsimiles that could easily pass as children's toys, to professional-grade gems painstakingly crafted from the finest materials. Some tools are made to last a lifetime or longer, others to be thrown away after a single use. There are strangely shaped tools with incredibly limited applications, and equally odd-looking varieties that, supposedly, can cut the weight of a carpenter's tool box in half.

Selecting a practical, reliable, and efficient set of construction tools from this abundant supply is analogous to ordering a meal from a four-page menu in an unfamiliar restaurant. The choices seem endless, and many of them are downright mysterious. You can't be sure about quality. You have no parameters for the prices. Will you get a lot or a little for your money? Is one item so crucial to the success of the whole that it warrants more investigation, and more of an investment than other, more peripheral goodies? How important is service and maintenance? Should you ask about ingredients, processing, packaging?

Faced with such a wide range of possibilities and questions, the first step is to narrow down the choices. To do this you should determine what tool-assisted jobs you want to accomplish, and what expectations you have about a tool's performance while you work. But it is not necessary to accumulate a basement full of tools before you build a deck or even a small addition. It is also unnecessary to buy the most impressive, most expensive version of everything in the store.

"Always buy top-quality tools." This easy answer is one more in an endless series of meaningless suggestions that have become standard clichés of do-it-yourself language. Obviously, if everyone could always buy the very best version of a product they would, and it would be nice. But buying the best, like many of the new how-to platitudes, is not always possible, practical, or, more importantly, even necessary.

The most sensible way to accumulate tools is gradually, as you need them, when the job is at hand. As you plan a specific piece of work you will become familiar with its general requirements and some of its subtleties, for example, the size of a chisel, and the type of circular saw blade that will work most efficiently. Long-range planning for tool and equipment requirements is more difficult but it can be done by examining every phase of a job carefully, including the materials used, and each step of construction.

As to cost, you should probably dismiss the very high and the very low. In a system rating the toylike tools at one, and the top of the line gems at ten, you should concentrate on tools and equipment in the five to eight range. As a general rule, an exceptionally cheap price reflects inadequate material content, and a shoddy, or at best, minimal manufacturing process. The top price generally reflects cosmetic features or bell-and-whistle extras that do not directly aid the performance of a tool's primary function.

It is also important to consider your expectations realistically, remembering that even the most sophisticated hand and power tools require an operator. And in almost every construction situation it is the operator who determines work quality, not the tool. In this regard, it does not make sense to buy tools that limit your skills. Once you have learned to cut a 2×4 in half leaving two truly square edges (and this is no small task), it is both frustrating and a waste of effort to use a flimsy handsaw with cutting teeth that won't hold an edge or a set.

Conversely, you don't need a blade of the finest Swedish steel, and a hand-carved, rosewood handle for a few minutes of rough cutting two or three times a year. Be honest with yourself about the work you have to do and the skills you have to do it with. Then buy your tools accordingly.

But (and this is a but that can mean a big difference in the amount of money you sink into tools and equipment) if you are developing building as a full-blown hobby, if you use tools on a fairly regular basis around the house (like every weekend), or if you need them for a major piece of prolonged construction (like an addition to your house), consider commercial-duty tools. The dime-store facsimiles, and even the low-cost, homeowner specials specifically targeted for light-duty work and do-it-yourself operators, will not stand up under production conditions approaching those encountered by professional carpenters.

Be sensible. Don't get sucked into a contest between the glittering appearance and collective price tag of your shop versus your neighbor's. High-quality tools are usually good looking, fascinating, useful, and expensive. But whatever their attraction, your overriding criteria in buying them should be necessity, and in particular, efficiency.

LENGTH OF SERVICE

In addition to its many positive traits, American industry is known for an anachronism called planned obsolescence. It makes little or no sense when money is tight, when competition among manufacturers is intense, and when products are being exposed to intelligent consumer scrutiny with increasing thoroughness. But it still exists in industry, and the tool industry is no exception.

Typically, this marketing gimmick builds one or more key elements with a specifically limited lifespan into the final product. In some cases this is inherent in the design. In others it is the fruit of subordinating quality to sales. You wind up with a five-year-old circular saw that is ready to run in every respect, except for the burned-out motor. This is an excellent reason for evaluating and comparing warranties carefully.

There are other surprisingly obvious twists to planned obsolescence, notably throw-away tools. Every hardware store and home center, and even many supermarkets, have bins of cheap tools. The idea behind this cornucopia of low ticket, impulse items is simple. Why spend ten dollars or more on a quality paint brush, for example, that is flagged and tipped to hold and spread paint efficiently, when you can get a facsimile for a dollar, use it once, and junk it. You can get a throw-away roller sleeve, and a uselessly thin and flimsy throw-away drop cloth too. It's the TV dinner approach to construction tools and equipment—no muss, no fuss, no cleanup, everything into the garbage.

Supposedly, it's the low price that makes these tools disposable, the idea being if it only costs a dollar, why keep it. But who throws away dollar bills, even one at a time? However, the two most prominent characteristics of throw-away tools are poor quality and inadequate durability. In other words, even if you tried to stretch that disposable dollar bill you couldn't, because the cheap tool it bought wouldn't stand up to a second use.

The cheap paint brush is a good example. The uneven, poor quality nylon bristles become waterlogged quickly; they don't hold paint well, which means more time and effort on the job, and if you try to clean them, many of the bristles come loose while those that remain swell and put the brush out of shape—altogether a losing proposition.

On the other hand, quality tools with good durability can last long enough, and provide enough service to pay for themselves. And quality is distinct from fancy, whiz bang extras. Every time you use a quality tool that is well matched to the job at hand, and to your proficiency as the operator, the work will be a little easier, and the results will be a little better. And if you maintain the interest, a good, commercial-duty tool will also last long enough for you to get good at using it to its capacity.

The signs of quality are listed here, not for every tool and piece of equipment, but for those commonly used in framing construction.

QUALITY CUTTING EDGES

The business ends of many essential construction tools are made from metal, particularly steel. But there are many distinctions among the four basic metals used in cutting tools: medium carbon steel, high carbon steel, high speed steel, and tungsten carbide.

Medium carbon steel contains 0.3 to 0.6 percent carbon. It is considerably stronger than malleable, low carbon steel (also called mild or machine steel), and may be used for bolts and some hand tools. High carbon steel (0.6 to 1.25 percent carbon content), also called tool steel, is commonly used for drills, taps and dies, cold chisels, and files. In high-speed steel, also called alloy steel, percentages of molybdenum and tungsten are added to help withstand the heat of friction produced by power sawing, drilling, and routing. Tungsten carbide alloy is the top of the line, consisting of particles bonded together at high temperature that approach the hardness of a diamond.

To set up a kind of test case where you could compare the different characteristics of these metals, picture an exposed ceiling beam made up of two Douglas fir 2 × 10's, surfaced on all sides with $\frac{1}{4}$-inch thick mahogany veneer plywood. Medium carbon steel blades would work well on the relatively soft structural core of the beam but would fracture the surface veneer, leaving ragged, uneven cut lines. Tools of high carbon steel, and even low alloy steel, might, if handled gingerly, leave the face of the veneer intact but would likely splinter the veneer on the back side of the beam as the blade passed through. A fine-toothed, tungsten carbide blade, however, would go through the veneer envelope and the structural core like a hot knife through butter, leaving finished, razor-edged cuts on both veneer surfaces of the beam.

HARDENING

While metal composition is important, the characteristics of each metal type can be altered by heat treating and tempering. The hardening process is accomplished in three stages. Initially, the steel is heated to about 1,600°F, then cooled (a process called quenching) in a bath of water, brine, or oil. Each type of steel has a dif-

Framing Tools and Equipment

ferent critical temperature at which structural changes occur (it varies according to carbon content). Heating beyond this point makes the steel fine-grained and extremely hard. It also makes it brittle.

This negative by-product of hardening can be overcome by tempering, or reheating at reduced temperatures below the critical change level, generally between 400° and 600°F. To harden a chisel, for instance, about an inch of the cutting tip should be heated with a blowtorch until the entire end turns a dark brown (this color indicates a temperature around 500°F) and then quenched. Long-handled pliers and a pair of gloves are essential safety measures for this operation.

The following chart provides color guides for tempering steel tools for specific purposes.

Tempering Color Guide

Color	Temperature (degrees F.)	Typical Tool Uses
Faint yellow	420	Hammers, cutting knives
Light yellow	440	Scrapers
Light tan	460	Dies, reamers
Light brown	480	Twist drills
Dark brown	500	Wood chisels
Purple	540	Cold chisels
Blue	560	Screwdrivers
Dark blue	600	Plane blades

METHODS OF MANUFACTURE

It is possible for excellent materials to be wasted in tools that are poorly manufactured, but it is unlikely. Quality materials almost invariably indicate quality tool construction, while low-grade steel content generally indicates inferior manufacturing. So it is usually possible to make an accurate judgment about overall tool quality based on the distinctions among manufacturing processes, particularly casting, machining, and forging.

When the bodies or housings of power tools, for instance, are cast, molten steel is poured into a mold of the tool's shape and allowed to cool. This process allows small bubbles to form in the steel, and this weakens it. A cast tool body may fracture if dropped, particularly in cold weather. And a cast wrench handle may simply snap under pressure. This low-quality process can often be recognized by rough edges of unfinished steel. This material is left over from the mold pour the way plastic parts of a model kit have a thin skirt of excess plastic around the edges that must be trimmed off before pieces can be fitted snugly.

Both machining and forging start with an ingot of steel that is heated, and then squeezed between heavy rollers to remove most of the weakening bubbles trapped as the ingot was formed. A machined tool is ground out of this rolled ingot like a piece of sculpture, often leaving telltale, symmetrical graining lines from the cutting or grinding tool. A forged tool is stamped out of the ingot with a huge, heavy, molded hammer under incredible pressure. This method further reduces the presence of bubbles and creates tools, generally referred to as drop-forged, that withstand great stress.

LAYOUT TOOLS

Many people invest most of their comparison shopping efforts and their money in tools like saws and hammers, which physically shape the structure, while neglecting planning and layout tools like rulers and levels. Certainly every tool and piece of equipment is important. If it isn't, if you don't depend on it for some part of the job, however small, it isn't worth buying. But layout tools are used to translate design ideas into physical relationships, and this gives them a special kind of importance.

The structural effects of sloppy sawing, hammering, and drilling can almost always be overcome. Building is an imperfect process replete with marginal inaccuracies and miscalculations. But a carpenter with a little imagination can find a way to overcome the most prominent mistakes without tearing things apart. The results may not be pretty but the structurally sound solution can be buried inside finished walls.

However, as in foundation work, inconsistent measurements and inaccurate layouts will show up in every step of construction, making each step more complicated and time consuming. A well-planned and executed layout is the glue that makes independent structural materials hold together in a building system. In addition, it is extremely difficult to compensate for bad planning in the final stages of construction because finishing materials can't be buried.

RULERS As noted on page 31, a 50- or 100-foot Mylar-coated steel layout tape is an essential tool for the initial stages of large-scale projects. Measuring long, house-size distances piecemeal, by 6- or 8-foot ruler lengths, inevitably creates inaccuracies and increases the possibility of a major error.

There are two basic types of short rulers: the retractable steel tape and the wooden folding extension rule. Each has advantages and disadvantages in different building situations. The steel tape is conveniently retractable, rolling up into a small case, and easy to carry. Specially clad versions will retain their imprint and resist rust indefinitely. The disadvantages, however, are numerous. It is hard to keep the tape rigid in free space measurements over four or five feet. And to make inside measurements (the distance between the inside edges of two studs, for example), the width of the case must be added to the dimension shown on the tape. This is a time-consuming and highly inaccurate procedure.

A wooden extension rule will stay rigid enough to measure 6- or 8-foot lengths in free space. The most useful versions have a thin brass 6-inch inset that slides out for easy and completely accurate inside measuring. Overall it is a bit more time consuming to use than a steel tape (all that folding and unfolding), and its imprint is less durable. A nice extra is a retractable

Framing Tools and Equipment

hook on one end of the wooden rule (you won't find this feature on dime-store versions), which is long enough to grab the edge of a stud securely, for instance, and which will not work loose and affect overall dimensions, unlike the rivet-attached hook on a steel tape.

Don't mistake this seemingly extreme attention to measurement and layout detail as needless nitpicking. Bear in mind that it doesn't matter how square, how plumb, how flawlessly accurate your cut is, if it's not in the right place. In this regard, and no matter what kind of ruler you decide to use, try to make as few independent measurements as possible. If you have to lay out and mark the locations of twelve joists across a beam, don't start at one end, measure out 16 inches, bring the end of the rule forward to this mark, and start from scratch again. It may take a little practice, but you will soon get used to the common 16-inch multiples of 32, 48, 64, and 80 inches.

SQUARES Even experienced carpenters use squares to mark true, right-angle lines before they start cutting. Three types are needed for framing work. A rafter square has an L-shape with one 24-inch leg and one $16\frac{1}{2}$ inch leg. It can serve as a straightedge to mark long cuts, and to mark square cuts into sheets of plywood. (In Chapter 11, on roof framing, you can find a key to the hieroglyphics etched into framing squares that some builders, who seem to think that houses work like mathematical equations, use to calculate stair layouts, rafter angles, and rafter tail cuts.)

A combination square has an angled handle with one straight 12-inch blade. It is used frequently to mark cuts, and by some to guide circular saws on dimensional lumber. A bevel square has a single, unmarked blade that can be swiveled to any degree; it is valuable since stair runners and rafter tails do not always conform to their well-planned mathematical models.

Both combination and rafter squares should have hash marks and numbers cut into the metal, not simply stamped on the surface. On combination squares there are notable distinctions to consider. Handle edges may be cast (cheaper but less true and durable), or machined (more expensive but more accurate). Don't pay extra for handles that hold single-vial spirit levels. They are too small to provide accurate readings.

LEVELS As noted in Chapter 3, an accurate level is indispensable. A level that is even marginally inaccurate is useless. If you try to level a 24-foot beam with a 2-foot level that's off only $\frac{1}{32}$ of an inch, that error, magnified twelve times in the 24-foot span, will be $\frac{3}{8}$ of an inch—too much for a critical beam. There are many types of levels, but only the first two listed here are practical and efficient for framing work.

• Water levels (see page 32) are inexpensive, easy to use, and completely reliable even over long distances, which is a unique combination.

• Carpenter's levels are the most practical for all around framing work. If possible, you should buy two: a four-footer for beams, joists, and studs; and a two-footer for door headers and other cramped spaces. Both versions have three bubble vials to read horizontal and vertical planes. Longer models, called mason's levels, are more accurate over house-size spans but are extremely expensive. Look for either a machined aluminum level, or a hardwood level with metal corner banding. The banding, which helps to preserve a true, straight edge, is important because these tools are used constantly to mark cut-offs on 4-foot plywood sheets. Some metal levels are available with removable bubble vials so you don't have to throw away the whole tool if only one vial breaks. Rough handling may not break a vial, but it can easily destroy the accuracy of these sensitive instruments.

Two levels that are not reliable are the torpedo level and the line level. Torpedo levels are shorter than others and contain one horizontal vial, one vertical vial, and sometimes a 45° vial. But the bubbles are too close together to provide accurate readings of most framing size dimensions. A line level is a small tube (about ½-inch diameter and 3 inches long) with a single horizontal bubble vial. The tube is hooked over a string, which can be stretched from one side of a foundation to the other, for instance. Readings from it are not reliable because they can be affected by string tension or even a slight breeze.

• Transits are very expensive tools that incorporate an extremely sensitive level with high-grade optics. This is a professional tool and is included here only as a sign of contractor quality. If you hire a guy to build a house or a large addition, and he starts off by setting up a transit, you are likely to be in competent hands.

The following process of elimination can be used to determine what level to use in different building situations. Long spans should be checked with the longest possible level, either a water level or a 4-foot carpenter's level. For shorter spans where a 4-footer won't fit, use a 2-foot carpenter's level. Under 2 feet it is more accurate to work by dimension, i.e., each side of the sill on a 1-foot-6-inch-wide window will be 20 inches off the floor. Given a relatively level floor, this method paves the way for the straightforward installation of finishing materials that will be cut and fitted by dimension, not by plumb and level lines.

CHALK LINES This inexpensive tool is simply a long string wound up in a box filled with colored chalk. It is essential for marking repetitive dimensions that are common in almost all framing systems. For example, if you measure in from the side of a building to mark both ends of a long stud wall, it is unnecessary to repeat this process for all the nailing points between these two points where the wall will cross over floor joists. The line, pulled tight, and held at the two end marks, can be snapped to deposit a neat, readable chalk line that is reliably straight. Don't try to use a chalk box as a plumb bob, even though some manufacturers bill this use as an extra feature.

PLUMB BOBS Described on page

Framing Tools and Equipment

34, this tool can be used to locate one point directly below, or plumb with, another—for instance, a point from the roof ridge to the first floor decking. It is accurate but time consuming to use. A much more practical facsimile is a very straight 2 × 4 of whatever dimension the situation calls for (placed with one end on the decking and one end against the roof ridge), which can be plumbed with a 4-foot carpenter's level.

CONSTRUCTION TOOLS

Saws

There are endless varieties to choose from, but almost all frame cutting can be done with a circular saw. This power tool gets a lot of use and abuse, so select it carefully. And as with all saws, the blade is a crucial consideration. A sharp blade can make a poor quality saw perform adequately. A dull one can make the best saw ineffective.

CIRCULAR SAW For general framing work use a 7½-inch model (that's the blade size). Smaller saws can't cut through large timbers in one pass, and are usually underpowered for construction work that involves some production cutting. Larger sizes (8¼ inches is the next step up) are likely to be unwieldy for most people but do increase cutting power and capacity. A motor smaller than 2 hp, a common characteristic of do-it-yourself saws, is fine for odd jobs, but not for repetitive cutting. Examine the following features carefully, and take the time to compare warranties.

- *Balance* The grip and trigger configuration must be comfortable and positioned for efficient leverage when you direct the saw through a cut. A handle high above or far forward of the motor housing makes for inefficient physics—you will use too much muscle, and may wind up dragging or pulling the saw.
- *Adjustments* The saw should have positive locking, easy to use guides that control the depth and angle of a cut. Large locking levers are best. Small wing nut locks are a cheap substitute and give the saw one more set of potentially dangerous edges. Pay attention to detail. It is unusual to find only one shabby feature on a power tool, the same way it is unusual to find only one sign of high quality.
- *Blade guard* The guard must cover the blade automatically after a cut is completed. A relatively large, easy-to-grip guard handle will make it possible to retract the guard safely at the beginning of mid-board plunge cuts.
- *Arbor lock* This feature enables you to unscrew the arbor nut that holds the blade in place without dangerously spinning the blade at the same time. It is a small but essential part of the saw's design.
- *Casing* As detailed on page 125, avoid cheap cast-metal housings. They just won't stand up to heavy-duty use. You are also likely to find many saws with housings made partially or entirely of plastic. Beware. They are advertised and justified as double-insulated protection against electrical shock. Fine. But all saws have three-wire, grounded cords to

prevent shocks. Additionally, the National Electrical Code now mandates ground fault circuit interrupters (special circuit breakers that trip at the most minor current leakage) for all exterior outlets that you are most likely to use when sawing.

The extra safety margin of double insulation should not obscure the fact that a quality metal housing is stronger and more durable than a plastic housing; but it is also more expensive to manufacture. An acceptable alternative is a sturdy, thick-walled, specially treated, impact-resistant plastic like Lexan. If the plastic is billed as impact-resistant but has the look and feel of a Cracker Jack prize, pass it up.

• *Automatic brake* This safety feature, which stops the blade quickly, and with a bit of a screech, when the trigger is released, is only valuable if you are likely to use the saw carelessly.

• *Rip fence* This feature is an afterthought on almost all circular saws. Its purpose is to guide the saw on a straight cut, say 2 inches in from the factory edge of a piece of plywood. Most have enough play in their adjustments and locking mechanisms to encourage blade binding for an inexperienced operator. Unless the fence is very sturdy, you are better off using a solid, 4-foot carpenter's level as a cutting guide.

• *Second handle* This is a nice but unessential extra. A front-mounted knob may help you guide the saw more accurately and securely when cutting studs, joists, or rafters in place. Don't use it to push the saw beyond its capacity. Let the motor do the work.

• *Blades* A combination blade, sometimes called a combination crosscut, is ideal for framing work. These blades should be made of tempered steel with 24 teeth that are taper ground to reduce friction. Taper grinding makes the cutting circumference of the blade slightly thicker than the center to minimize resistance as the heart of the blade passes through the cut. High-quality steel should keep an edge for several weeks of production cutting. But if you hit a nail by accident, or cut through knot-laden wood, the teeth can be touched up with a mill bastard file. Blades with tungsten-carbide tips are much more expensive but stay incredibly sharp for a long time. However, they must be professionally sharpened.

To get the most out of saw and blade, always wait until the maximum rpm level is reached (this takes only a fraction of a second) before starting the cut. All these characteristics and features of circular saws should be carefully considered because this tool is the primary power extension of your hand during framing work.

CROSSCUT SAW It is impractical to use a power saw in every building situation. And the hand-powered substitute is a crosscut saw. To provide good service the saw should have a comfortable, hardwood handle, secured to the blade with five rivet screws. The blade, commonly 26 inches long, should be tempered and taper ground. Crosscut saws are also

Framing Tools and Equipment

called 8-point saws for the number of cutting teeth per inch. Saws with fewer teeth (5 per inch) are for rip cuts, or cuts in line with the wood grain. More teeth (10 or 12 per inch) are for fine cuts and trim work.

HACKSAW Primarily an electrician's tool, this saw is a mistake fixer for carpenters. On framing, it is used to cut through nail shanks where the heads have been driven or set too deeply to be pulled out, to cut through lag bolts that can't be tightened or removed because of stripped threads or heads, and to customize frame hanging hardware in special situations.

MISCELLANEOUS SAWS One nice extra is a stubby crosscut saw for trimming assembled timbers in hard to reach places. But a good 16-inch or so model is difficult to find. Similar odd, small-space jobs can be handled with a power reciprocating saw (not a vertical blade jigsaw), although this tool provides all the finesse of a meat cleaver.

You do not need the following saws and accessories for general framing work: a rip or trim saw, coping saw, jigsaw, compass or keyhole saw, back saw, or miter box.

Hammers

A hammer is the most personal framing tool because it is an immediate, hand-powered extension of your body. The key to quality in a hammer is a high-grade steel head that is drop-forged. This minimizes the possibility of chipping the head with off-center hits on steel-capped wood chisels, on cold chisels, and particularly on cut or case-hardened nails.

Handles may be hardwood, reinforced fiberglass, or steel with a rubber sheath. Fiberglass handles will occasionally shear up near the head, and this is an abrupt, unpleasant experience. Rubber-cushioned steel is certainly solid but a bit bouncy. Hardwood strikes the best balance between strength and resilient shock absorption, but, if it isn't obvious already, this is purely a matter of personal preference.

A 16-ounce hammer is minimal. Less weight in the head makes you provide most of the driving force. More weight (20 ounces is the next common size) lets the hammer do more of the work, but may take some getting used to. If you don't work with tools on a regular basis, four hours of nailing off plywood roof decking with a 20-ounce hammer will make your hand and forearm feel like a wet noodle.

Roman claw hammers (the nail pulling end is gently curved) are the most practical and the safest. Larger, 28-ounce, straight claw, ripping hammers are monsters reserved for heavy timber construction where 6 × 6's and 8 × 8's are set with 60d (or 60-penny, that is 6-inch) nails.

A hammer holster is an invaluable accessory, not a do-it-yourself affectation. It consists of a horizontal steel ring protruding from a piece of leather that can be fitted on your belt. Removing and sheathing the hammer is easy if the horizontal ring is large and sturdy. In construction there are many operations that require two hands. And fiddling with a small cloth loop attached to the leg of your dungarees, or constantly searching for a safe and convenient place where you can park your hammer, are disrupting and time-wasting procedures.

SLEDGE HAMMERS Known affectionately in the trade as a persuader, a 10- or 12-pound, drop-forged sledge provides the extra punch needed to move seemingly immovable objects. A case in point: You assemble a stud wall in strict accordance with a carefully planned layout, yet a final check shows one corner post a full ¼-inch out of plumb. And this is just the kind of baffling development that arises in house construction. You try adding a few nails to draw the post into line, with no success. But one good shot with the persuader does the job.

Remember, never hit a hammer with another hammer. And don't use the claw end of a hammer to do the job of a 5-pound crowbar. While it is unnecessary to own one of everything in the hardware store, it is inefficient and sometimes dangerous to use tools for purposes beyond their designs.

Levers

These tools are not directly used to build anything. But the highly imperfect process of construction requires them for wedging, lifting, twisting, pulling nails, and other jobs associated with real as opposed to ideal conditions. Two varieties will come in handy.

CROWBARS Also called pry bars and ripping bars, these tools are designed to give you leverage where your strength is not sufficient or where no hand holds are available. For example, in laying 2 × 4 sleepers (flat side down) over an irregular concrete slab, one of the 2 × 4's may be nailed down into a depression. There is no way you can pull it back up, to make it level with the other sleepers, by hand. But a pry bar, slipped between concrete and sleeper, applies more than enough force. An 18- to 24-inch, drop-forged, high carbon steel ripping bar is about the most versatile design. One end is crooked, the other is flat. Both prying ends have a nail-pulling jaw. Larger bars are available, all the way up to 10-foot wrecking bars that are a struggle to get off the ground.

NAIL PULLERS This smaller, more subtle version of a crowbar should also be drop-forged, high quality steel. Length is about 12 inches, with one end cupped into a miniature, concave claw that gives this tool its slang name of cat's paw. This shape, with some practice, can be driven under the heads of deeply driven common nails without gouging the face of the lumber.

Drills

There are not a lot of jobs for a drill in framing work. Consequently, you may get by with a do-it-yourself, ¼-inch drill for occasional pilot holes and other odd jobs. But you would only be getting by. In later stages of construction you will need a rugged drill that can easily go through structural members to provide access for piping and electrical cables. Also, framing systems that use 4-inch-thick dimen-

Framing Tools and Equipment

sional timbers require pilot drilling and large counterbores to make room for lag bolts and their seating washers.

A $3/8$-inch reversible drill with adjustable speed control and a $1/3$ or larger horsepower motor, is a good, all around selection. Look at extra features carefully. An auxiliary handle, for instance, may seem like a nice extra, but it makes it easy to apply more pressure on the drill than the motor can cope with. But some extras can be valuable: for example, a rubber chuck key holder that wraps around the drill cord to keep the chuck key just out of the way but always at hand.

Two notes of caution: first, never move the reversing lever while the drill is spinning; second, do not use the button (if provided by the manufacturer—and many of them still include it) that locks the trigger in an on position. This is a dangerous feature, particularly on a heavy-duty drill. A powerful $3/8$- or $1/2$-inch, triple-geared drill can literally toss you off your feet when the bit locks in a knot, and the drill body starts to turn instead of the bit.

BITS Start with a simple collection of common sizes: for small hole drilling, twist drill bits from $1/16$- to $1/2$-inch by $1/16$-inch increments (eight total); for drilling in concrete or block, masonry bits (carbide-tipped bits are more expensive but cut quicker and last longer), from $1/4$- to $3/4$-inch by $1/8$-inch increments (five total); for larger holes in wood, wood-boring (also called flathead) bits from $1/4$- to $1 1/4$ inch by $1/8$-inch increments (nine total).

Wood boring bits can be touched up with a file to produce cleaner cuts. For exposed counterbores, Forstner bits with a circular cutter surrounding the starter point provide an exceptionally clean-cutting but expensive alternative to flathead bits.

Planes

A small, adjustable, sharp, block plane is essential for framing work. It is the tool that literally smooths out most of the rough edges between realistic field conditions and idealistic textbook conditions. Tools like these are crucial because construction is not mathematically perfect. Cuts are rarely true. Occasionally you will hit one right on the nose: You trim a 4×4 post, plumb it on a 2×4 shoe, and find that all four sides of the joint meet in a fine hairline seam as tightly as the joint on a fine hardwood cabinet.

Conversely, sometimes a cut will be so far out of square or out of plumb that the board becomes useless, although you can always turn it into pieces of blocking or bridging. But most cuts are somewhere between these extremes—just a little ragged, a little out of plumb, not enough to decrease substantially the strength of the joint, but enough to make them obviously imperfect. And a sharp, heavy-duty block plane, applied with a firm and controlled hand, can turn most imperfect cuts, most slightly cupped or twisted boards, most timbers with knots in the wrong places, and most joists and rafters with rough, splintering chamfers, into quality pieces of construction.

Block planes are one-hand tools. They

are light (about 1 pound, 8 ounces), 6 to 7 inches long, usually about $1\frac{5}{8}$ inches wide, with a blade set at a 15 to 20° angle so you can cut with or across the grain. The plane should have solid, fine-scale adjustments for aligning the cutter blade with the plane body laterally, for controlling cutter protrusion (depth of cut) below the plane bed, and, on some models, for opening and closing the mouth in front of the blade where shavings are released.

Jack planes are larger, 14 inches long with a 30° to 35° bevel on the blade, which is placed at a steep angle in the plane approaching 45°. It is useful for removing more wood in less time. A jack plane is too bulky for cleaning work on cut ends of timbers and too gross to cut end grain unless the blade is exceptionally sharp. It is ideal for reducing crowns, or high spots, on particularly convex joists and rafters. And like the block plane it is invaluable for work in latter stages of construction, particularly planing doors and frames around door and window openings.

Chisels

Occasionally you will have to make a mortise, or pocket cut, in timbers that can be accomplished only with hammer and chisel. For framing, use chisels that can withstand heavy-duty use, particularly repetitive hammer blows. A set of butt chisels, from $\frac{1}{4}$- to $1\frac{1}{4}$-inch widths, by $\frac{1}{4}$-inch increments, is more than adequate.

Plastic handles (make sure they are made of high-impact plastic) should be topped with steel striking caps. Sharp chisels are convenient for cleaning tenon cuts, half laps, and many other odd jobs where even a small block plane is too cumbersome or can't get in tight enough to make an inside, right-angle corner.

The cutting ends of chisels should be razor sharp so the tool can do the cutting while you provide direction and control. Preparing a tool for work (sharpening in particular) can sometimes take more time and effort than the work itself. But this lopsided ratio of preparation to production makes sense. Just think of the difference between shaving with an old, dull razor blade and a brand-new one.

With almost all hand tools there is a proportion of brain and brawn that goes into the operation. On cutting tools you can transfer brawn to the tool more efficiently by increasing its sharpness. This frees you to pay more attention to details and overall accuracy.

Clamps

You can find a clamp for every situation. There are literally hundreds of specialized varieties. But for framing work, three types (two of each is sufficient to start with) will cover all eventualities.

Spring clamps (either 6 inches long, opening to 2 inches, or 9 inches long, opening to 3 inches) work like clothespins. They provide a third holding hand on small cleats, plywood fillers and gusset plates, joist hangers, and the like, when you need one hand for the hammer and one to hold the nail.

Framing Tools and Equipment

Bar clamps (24 inches long with a 2- or 4-inch throat capacity) are large enough to hold built-up structural beams together during gluing and nailing. They are also strong enough to provide constant leverage—again when you can't spare one hand for a pry bar. A case in point. You assemble a stud wall, 16 inches on center, which should make the distance between studs 14½ inches. But before cutting a stiffening cleat to fill this space, you find that one stud is bowed enough to make this distance 15¼ inches instead. Should you cut an odd-sized cleat to fit? No. Why make the deficiency in the board permanent? A bar clamp has the torque to remove the bow so you can fit in the proper-sized cleat and nail it home.

Pipe clamps can be used for the same purpose on larger timbers over longer distances. The actual clamp fittings are bought in pairs, independently of as many different lengths as you might need of standard, ¾-inch, black iron pipe. One half of the clamp, a viselike fitting, is screwed on to the threaded end of the pipe. The other half of the clamp can be locked and released with a series of spring-loaded cams, so it can be moved to any point on the pipe.

You can buy two sets of clamp fittings, and different pairs of pipe from 2-feet to 12-feet or larger. If oiled regularly, the fittings can be easily changed from one set of pipes to another. These clamps are useful for aligning bowed studs, corner posts, and rafters, particularly 4-inch wide dimensional timbers that are a bit much to unbow by hand.

Miscellaneous Hand Tools

It is sensible to acquire a standard collection of hand tools (you may already have most of them), including a caulking gun that accepts tube-packaged construction adhesive, and things like screwdrivers (flat head and Phillips), files, rasps, pliers, and a variety of wrenches. One type, a ¼- or ⅜-inch drive socket wrench, is used frequently on large timber construction.

On framing systems that use 4-inch or larger timbers, connections commonly made with nails can be secured more solidly with lag bolts. This also holds true at critical stress connections on 2-inch dimensional systems, for instance, where the ledger supporting joists of a deck is attached to the house frame. Lagging requires drilling pilot holes and counterbores deep enough to recess the washer and bolt head.

A common open-end wrench cannot be used in this hole so the lag must be turned with a socket wrench. Remember, though, that you need only the socket sizes that correspond to the lag bolt heads, not a glittering, umpteen-piece, car mechanic's set.

Air Tools

These tools deserve only a quick note as their use is really limited to commercial, production building. Simply, this group of power tools (everything from staplers used on asphalt strip shingles, to hammers, drills, and planes) run on compressed air instead of electricity.

However, in new construction, even on

a relatively remote, uncleared site, it is customary, and not very hard, to have the local utility company run in a temporary electrical service before construction begins. In extremely remote sites, gasoline-fueled generators can be used to provide power, or gasoline-fueled air compressors. Air-powered tools are specialized and cannot be adapted to household current.

CONSTRUCTION EQUIPMENT

Ladders

On most addition-size or larger projects you will find many uses for a 6- or 8-foot stepladder. On some jobs, 16-foot or longer extension ladders are needed. For both types, heavy-duty, aluminum ladders make the most sense because they last longer than wooden ladders, they can be exposed to the elements without deteriorating, and particularly because the signs of structural weakness are easy to recognize in metal, but not in wood.

Wooden ladders can swell, split, and work loose at critical joints in ways that are apparent only under close inspection. But aluminum ladders, which are flexible enough to absorb a lot of tension, show these deformations of excessive stress as deep crimps or tears in the metal that are easy to spot. If you do use a wooden ladder, lay it flat on the ground and walk along the rungs to test their strength before you go up in the air.

The feet of a ladder must be secure. On a wooden deck, you can temporarily nail a 2×4 cleat behind the feet to keep the ladder from sliding. On dirt, you can embed the feet in the earth and sink a 2×4 stake against the bottom step for extra security. In addition to common sense, and a realistic fear of heights (if you fall, it will hurt), the most basic safety rule is to keep your hips, and therefore most of your weight, within the borders of the ladder.

Extension ladders, in pairs, can also be used to provide a flexible scaffold system with the aid of ladder jacks attached to the rungs. These jacks can be adjusted to form a horizontal projection on which wooden scaffold planks can be laid.

Scaffolding

The standard scaffold plank is 2 inches thick, 9 inches wide, and 13 feet long. And in this special case the nominal dimensions are the same as the actual dimensions. A Douglas fir 2×10 (the actual dimension is reduced to $1\frac{1}{2} \times 9\frac{1}{2}$ inches) should not be laid on the flat and used as scaffolding, particularly if it contains large, full-thickness knots.

You can rent planking, but if you do, inspect it carefully for knots and splits, and never use scaffolding that has been patched with nails, wooden cleats, or metal mending plates. Even if planking is sound, be sensible about loading it up. Deflection, or downward bending at midspan, in most structural beams, girders, and joists, is limited by design to $\frac{1}{360}$ of the span. In some cases this margin is reduced to $\frac{1}{180}$ of the span, or less than 1 inch over a 13-foot plank.

Remembering these structural guide-

Framing Tools and Equipment

lines, don't expect a scaffold plank, even at full 2-inch thickness, to support two 175-pound workers, their tools, and five 40-pound bundles of shingles at midspan. Concentrate a reasonable load as equally as possible over the two points where the plank is supported.

Two types of scaffolding equipment are practical for more than single-story construction, although they are used primarily for exterior finishing work like roofing and siding. Ladder jacks, used with a pair of extension ladders, can be hooked onto the ladder rungs and then adjusted so that a horizontal steel projection can securely support planking. These jacks can be fitted to project away from the building, on the outside of the ladder, or toward the building, on the inside of the ladder.

Jacks can be placed one at a time by one person, but two people are needed to bring the plank up and set it in place. When jacks are projected out from the ladders, away from the building, it is difficult to get around them from the rungs to the planks. These operations cannot be performed too carefully.

Pump jacks, used primarily for painting and applying siding on high, vertical walls, ride up and down on two vertical 2 × 4's (spiked together) that are staked at ground level and braced to the building on top with metal standoffs. A friction roller cam, pumped with your foot, raises and lowers the support carriage carrying the planks. Elaborate accessories are available, including projections for tool holders, materials, and a guard rail.

Sawhorses

Most construction-size timbers are too large and too heavy to hold, measure, and cut in one hand with any margin of safety. Similarly, you can't drape yourself over the top steps of an extension ladder to trim rafter tails, or cut timbers that are laid flat on the ground, or even on a few rickety blocks. It's not safe, and you won't get an accurate cut.

Your feet must be firmly planted, and the timbers you cut must be secure and stable. A solid pair of sawhorses satisfies these conditions, and gets the wood up to a convenient cutting level. You can make them from scratch out of 2 × 4's, cutting severe angles on the splayed legs where they join the main, horizontal 2 × 4 crosspiece. This V-shape should be reinforced with a plywood gusset plate covering the top 8 inches of the legs and nailed in at 2-inch intervals.

However, there is a common problem with sawhorses, homemade or store-bought. After you have loaded them up with timbers, and moved them around the job site, and laid planks across them to make scaffolding, they start to get wobbly. Most of this deterioration can be avoided by using aluminum sawhorse brackets. They are preformed to accept square-cut legs and a 2 × 4 crosspiece. (You don't have to cut oblique angles on the legs, which are the weakest point in the horse.) The legs can be fastened permanently with 12d common nails driven through the bracket, the 2 × 4, and then clinched (bent over), while the crosspiece is tacked temporarily. When you remove it, the legs fold together for easy storage.

If the horses are left outside, soak the end grain on the legs in a bucket of creosote or clear wood preservative and brush a liberal coat of preservative over all the wood surfaces. Although more elaborate, and more expensive, all-metal sawhorses are available, they don't provide special or improved service over wooden horses to make them worth the money.

Generators

In remote sites, or to get work started before the local utility company can run in a temporary electrical service, a gasoline-fueled generator can provide power for construction tools. They are expensive (over $1,000 for a 3,500- to 3,750-watt machine), so unless you are going into the construction business, or the building site can't be electrified, rent one if you need it. The monthly rates generally approach $200 for a 3,500-watt generator, and approximately $75 for one week, although short-term rental rates vary widely.

Most construction generators have recoil (pull rope) starters, with battery-powered electronic starting as an expensive option. Many provide dual voltage (120 and 240 volts), with multiple receptacles for extension cord plugs. Check for an overload indicator, and surge capacity (about 25 percent), to accommodate load-draining starts.

At full load, one gallon of fuel should last about an hour, and roughly twice as long at one quarter of the machine's electrical capacity. Long-run models that can provide power for 6 to 8 hours at full electrical load are equipped with an auxiliary fuel tank, fuel pump, and selector valve to switch between tanks.

If you use a generator approaching this size on a building site surrounded by existing residences, be prepared to run it only during normal construction hours (8 A.M. to 4 or 5 P.M., Monday through Saturday) as it makes as much noise as a small tank.

Extension Cords

Extension cords must be grounded (three-wire cable with a three-prong plug) and insulated. It makes no sense to defeat the built-in shock protection of a grounded power tool by plugging it into an old-fashioned, ungrounded, two-wire extension cord. Use cords that are at least 12-gauge, compared to house wiring, which has 14-gauge conductors, a smaller diameter wire.

Insulation is a must for extension cords used in cold weather, where a conventional plastic sheathing becomes quite stiff and unmanageable. Additional insulation will make the cord bulkier, but it will keep it supple enough to allow safe, unrestricted use of power tools.

Extremely long cords can be a nuisance (and sometimes a safety hazard) in the labyrinth of exposed framing members. And if the long cords are undersized they may cause a voltage drop that can damage power tools by straining their motors.

One efficient alternative is to make yourself a main extension cord of 10-gauge conductors that runs from the temporary power service provided by the utility company to the heart of the job site. At the job end, instead of a single plug, you can attach a conventional outlet box with two grounded duplex receptacles. Short extension cords to serve individual power tools can then be plugged into four grounded outlets in this central, heavy-duty line.

TABLE SAWS AND RADIAL ARM SAWS

These tools are listed together because they perform similar functions. They are listed at the end of the chapter because all the cuts they can make on framing materials are well within the capacity of a good circular saw. Since table and radial saws can do a lot more than cut framing timbers, and since they are large, heavy, and expensive, they are more suited for home shops than for job sites.

On table saws, wood is guided through a fixed-position blade, which can be adjusted only for depth and angle of cut. And on one variety of table saw, called a tilting arbor saw, cutting angles are changed by tilting the saw table. In contrast to a fixed blade both the motor and blade of a radial arm saw move on a track for cross cuts, while the wood is stationary. Consequently, long, dimensional timbers are difficult to cross cut on a table saw. Even a 10-foot 2 × 10 is a bit tail heavy to guide through the blade at right angles. But the same board is easily cross cut with one pull on a radial arm saw.

Conversely, ripping cuts (with the grain), and long cuts on wide sheets of plywood, are much easier to make on a table saw than on a radial arm saw where the cutting head must be rotated and locked. Thinking past the limited use on framing, both saws can be equipped with attachments to perform routing, dadoing, sanding, and many other woodworking operations.

TABLE SAW Also called a bench saw or a tilting arbor saw, sizes range from 8, 10, and 12 inches according to blade diameter. A top of the line, professional grade saw would be a 12-inch, $3\frac{1}{2}$ hp, 220-volt model that would plow through 4-inch hardwood stock like a fish through water. At the other end of the line you can get a surprisingly inexpensive, somewhat tinny, 8-inch, $\frac{1}{2}$ hp model that is suited only for home hobby use. A solid intermediate choice for reliable, fairly regular duty is a 10-inch model with a $1\frac{1}{4}$ hp motor, developing 3,450 rpm.

Two types of motors are available. Less expensive models use a universal motor that is connected to the blade spindle with a belt, relatively noisy, but cheaper than the second type, an induction motor. This is more expensive, quieter, and requires less maintenance because there are no carbon brushes or commutator and no drive belt. Be sure you know if a motor is included in the price of the saw. Frequently, it is not.

RADIAL ARM SAW Sizes range from 8 to 16 inches and even larger for some commercial models. The blade generally bolts directly to the motor shaft, eliminating the need for gears and belts.

A 12-inch, 1½ hp model developing 4,200 rpm should efficiently handle all framing and general woodworking requirements. The chief limiting factor on a radial arm saw is the maximum allowable rip width, or how far away from the rip fence the motor and blade head can be pulled and rotated to cut in line with the grain. If it is less than 24 inches, you will never be able to rip a 48-inch-wide sheet of plywood in half.

In addition to carefully examining and comparing warranties, you can effectively judge overall quality on table and radial saws with a hands-on test. Use every adjustment on the saw, including depth of cut, angle of cut, width of cut (closely check the positive lock and parallel alignment of the rip fence), and the overall construction. If you can force a rip fence out of parallel alignment with the blade using only a little pressure, you will have problems making accurate rip cuts.

Choose these and all your tools with care. Look hard for signs of quality, and don't be distracted by bell-and-whistle extras. Finally, remember that tools are inanimate objects. They don't have a mind of their own; they can only provide efficient, accurate, and reliable service under your control. Tools don't build houses— people do.

9

PROPERTIES AND DESIGN VALUES OF FRAMING MEMBERS

Wood comes in different species, grades, and sizes. This chapter explains the considerable distinctions between one piece of wood and another and how to choose the right combination of characteristics for any structural timber in a residential frame.

To be honest with you, this is quite a task, and at first glance a lot of the information may seem too technical to deal with. Many of the technicalities could have been buried in charts and tables at the end of the book, without explanation. Instead, all the figures, formulas, and relationships you need to determine what timber to use in a 6-foot porch floor or a 20-foot roof are integrated with explanations of lumber characteristics and design applications. This way the job of selecting the proper framing timber is reduced to the simple operation of plucking the right numbers from a framing design table.

If you looked ahead in this chapter and saw the pages of tables, you are probably asking yourself, "Are all these facts and figures really necessary?" Yes. And here's why. We are no longer building log cabins. If we were, and were cutting 12- or 16-inch-diameter trees to make massive joists, rafters, and walls, we would not have to be concerned with a matter of inches.

But using these solid, 16-inch-diameter trees would make your house tremendously overbuilt, i.e., many times stronger, safer, and more durable than is necessary. That might sound ideal to you, as the individual owner—after all, strength, safety, and durability are desirable characteristics in a house. But overbuilding doesn't make sense for all homeowners as a group, as standard operating procedure—not anymore.

Severe overbuilding is an economic sin and an ecological disaster. Just one of those 16-inch-diameter trees can provide joists for the entire first floor of a typical timber frame house, another, all the studs for the walls. Wood is a renewable resource, but growing trees takes time. And this natural growth has not kept pace with the increased demand for wood to be used as pulp for papers and magazines, for bags and boxes and endless amounts of packaging, and as lumber for construction.

During the Viet Nam War, for example, consumption of plywood by the military created severe shortages, and tripled the price of plywood sheathing used in house construction. When any material is in demand it gets expensive. And as the demand and price increase, the inclination is to use the material as sparingly as possible.

While this stinginess makes sense economically and ecologically, it may not always make sense structurally. Here's the problem. Selecting the right timbers for the right locations in a frame is a process influenced by cost, availability, building codes, architectural design, maintenance, attached materials, the type of structural system used, and other factors. Economy is important but it cannot be allowed to override all other considerations.

Like special interest lobbies in Washington, each factor wants a different response: for cost, the less wood the better; for maintenance, more wood for more strength, which means less structural shifting that can crack tile joints and pop nailheads in Sheetrock walls. Trying to satisfy every factor a little is as unprofitable as trying to satisfy only one factor completely.

The trick is to balance the demands of these different factors. This is the practical, logical way houses should be designed and built. But if piecing together this jigsaw puzzle seems too complicated for you, don't worry. Almost no one selects lumber this way because there are several other less complicated options. For instance, you can take an educated guess at rafter and joist sizes, and let the building inspector who checks your plans change the 2×6's to 2×8's. Of course, your guess may be so far off that the inspector realizes you don't know what you're doing, in which case your building permit may not be approved.

Alternatively, you can estimate timber properties and sizes based on experience, that is, if you have any. If the deck on your last house seemed pretty secure with 2×8 joists, why not use them again?

Or finally, whether you have construction experience or not, whether you are building a mansion or a set of porch steps, you can eliminate guesswork by using the selection of standardized lumber design value and span tables given here.

These tables are by far the best way to determine what species, what grade, and what size lumber to use in different parts of the frame. In fact, practicing architects and engineers use them as a matter of course. They have to rely on them because there is no single rule for integrating the elements of timber species, grade, and size into a structural design. Why? Because every wood species (pine, redwood, fir, etc.), and every grade (select, construction, utility, etc.) has its own set of characteristics.

Properties and Design Values of Framing Members

If you look at stacks of timbers in a lumberyard, or even a pile of firewood, it is evident that some wood is hard and some soft, some clear and straight grained, and some filled with knots and voids. The variations are endless.

Do these characteristics effect structural quality? Certainly. As an admittedly exaggerated example, it would be as ridiculous to build the main girder of a house out of balsa wood as it would be to build a toy glider out of knot-laden oak. But these distinctions are far more subtle in residential frame design.

Before plunging into the tables you should be aware of these subtleties, and the limitations of dry, textbook figures. While the tables are standardized and relatively easy to use once you understand the terminology and the ground rules, the information they contain should be tempered with experience, or at least common sense. For instance, if you plan to build a small den using 2 × 6 floor joists, but you want to install an immense fish tank in the middle of the room, try changing the 2 × 6's to 2 × 10's. And for precise load calculations in special situations, you should always seek professional help.

In essence, that is the case for minimal use of raw materials, and for taking the trouble to split hairs with detailed construction tables. From the thick, wasteful, uneconomical logs of the cabin we went to dimensional lumber (2 × 4's and 2 × 12's) for studs, joists, and rafters as determined by mathematical tables. Then we tempered this information with experience, or lacking it, common sense.

Now more confusion. Experience and common sense will also show that all the figures on the tables are conservative. In fact, they will dictate grades and sizes of wood that provide more strength than seems necessary. So are we back to the wasteful, costly, overbuilding syndrome again? Not really. Well, can a wall be made of 1 × 2's instead of 2 × 4's? Yes. Will it stand up? Yes. Will it blow over in a storm? Probably. And that's where the conservative values of the tables come in. They will dictate more strength than seems necessary, but not more strength than seems sensible. There's a big difference.

So the question becomes how much fat can you trim off the frame without undermining a reasonable margin of safety. This is one of the quagmires of residential construction today. It is a gray area complicated by considerations of material cost, labor time, maintenance requirements, and the durability of an overbuilt versus an underbuilt frame.

It would be rare to buy a house with floor joists so undersized that you fell through them into the cellar, with rafters so skimpy that a 6-inch snowfall collapsed the roof. The strength of framing timbers, even if grossly undersized, is enhanced when the timbers are tied together with other materials in a framing system. This characteristic provides even more of a safe design cushion. But is the cushion too big, too costly, and needlessly overprotective?

Well, lumber design value tables do

take this strength-enhancing factor into account. For instance, for each species and grade, the design value tables show two figures for fiber stress in bending, which is one of the characteristics we will decipher shortly. One bending value indicates the timber capacity when used alone, for example, as a single girder supporting floor joists. The second bending value indicates the timber capacity when used in combination with two or more similar framing members and a load distributing material, for example, as a series of floor joists tied together with plywood subflooring.

The tables allow a 15 percent decrease in strength requirements for this multiple use. In other words, there is strength in numbers, even extra strength, if the numbers are strung together correctly. But experience shows that this decrease is unrealistically low.

Ripping out a stud wall, even if it is buckling under the weight of shingles from many reroofing jobs, is no easy task. More often than not, the roof will continue to hang by a thread as you rip out stud after stud. In conventionally designed residential structures you can, in fact, remove every other stud, joist, and rafter, without causing the building to collapse. But when you walk across the floor it will bounce, and when you slam the front door the walls will quiver like the surface of a water bed.

So when experience and common sense demonstrate that the tables are conservative, and that timbers in combination have added strength, why worry about a difference of only a few inches in span length limitation that says your 2×6 should be a 2×8?

The answer lies in that gray area between design and construction; it harks back to the differences between minimally adequate structures and sound structures and to the process of balancing the demands of strength with material cost, of durability with inexpensive labor, and of long-term quality with short-term profits.

The tables, and building codes in general, contain a reasonable safety margin that, coupled with the enhanced strength of wood in a framing system, is able to accommodate honest mistakes in judgment, hidden defects in material, and even some slipshod methods of construction.

But as economic pressure tempts developers and builders to trim framing fat closer and closer to the bone, these margins of strength and safety become increasingly important. In fact, they are largely responsible for preserving what is left of housing durability. When a frame is only minimally strong, it does not do a good job of preserving the materials that are attached to it. Plaster, for instance, is rarely used on residential walls and ceilings anymore, partly due to its high cost, but mainly due to the frame strength needed to prevent cracking and continual maintenance.

It is an imprecise balancing act that requires sensible compromising. The word "sensible" to many will seem oddly out of place in this technical realm of specific mathematical building relationships. Yet

Properties and Design Values of Framing Members

it accurately represents the balance you should strive for in frame design.

So in sum, the best course is not the overbuilt, log cabin approach, and not the minimal, cut-every-corner approach. It is between these extremes in the lumber design value tables, and the span tables, cushioned by their built-in safety and strength factors, the desirability of a durable product, and your common sense.

FRAMING TIMBER SIZES

Most consumers know that you don't always get what you pay for. With many products this discrepancy is hidden away in the fine print of a warranty or buried beneath a slick exterior in a cheap, poorly manufactured motor. In the lumber business it is right out in the open.

The 2×4 studs in an old house measure 2×4 inches, as you would expect. In a house roughly ten to fifteen years old the studs will measure only $1\frac{5}{8} \times 3\frac{5}{8}$ inches due to a trimming process at the sawmill called dressing. Now, and for the last few years, that dressed stud is down to $1\frac{1}{2} \times 3\frac{1}{2}$ inches.

This example of fat cutting must be accounted for when you lay out framing timbers, and when you order materials. The specifications in design value and span tables do take this discrepancy between title and content into consideration. But you should be aware of these contradictions when you deal with a lumberyard.

Their customs include one name for the original size of a board (called the nominal dimension), and another name for its true size (called the actual dimension), a method for pricing some lumber by length and some by square footage. Here are the translations of lumberyard language you'll need.

NOMINAL These dimensions are used by lumberyards and on building plans for convenience. They indicate the size of wood before it is dried (either in a kiln or by a natural seasoning), and before it is dressed (trimmed to consistent width and depth dimensions at the mill). Undressed lumber, sometimes graded "Rough," may measure very close to the nominal dimensions.

ACTUAL These dimensions indicate true size. While joists in an old house may actually measure 2×10 inches (the current nominal size), those in a new house measure $1\frac{1}{2} \times 9\frac{1}{2}$ inches. An occasional load from the sawmill may come through $\frac{1}{16}$ inch larger each way. On a 24-foot long floor, requiring nineteen joists spaced 16 inches on center, this small, $\frac{1}{16}$-inch difference can alter the overall length of a layout marked for standard, $1\frac{1}{2}$-inch-wide joists by $1\frac{3}{16}$ inches, or nearly the width of another full joist. It pays to spot-check width and depth consistency at the yard.

GREEN Lumber that still contains free water in the wood cavities is green. When the cavities dry, the wood reaches what is called its fiber saturation point. At this level overall moisture content is still 25 to 30 percent. Past this point, continued drying pulls water from the individual cell walls in the wood instead of the open cavities between wood fibers. This kind of cell wall drying, below 25 or 30 percent, causes shrinkage.

SEASONED In order to avoid the warping and twisting that can occur as wood dries, water must be removed by seasoning or kiln drying before the lumber is placed into a framing system. Acceptable dry or seasoned lumber has a moisture content (the weight of water in wood compared to the weight of the dry wood only), of 19 percent or less. Some special grades are marked MC15, indicating a moisture content of 15 percent or less.

LINEAR FOOT Also called a running foot, this is the pricing measure used in lumberyards for some nondimensional (nonstructural) lumber like molding and dowels. Width and depth are not considered in the cost calculation.

BOARD FOOT This lumber measure is also used for pricing, and it is the most frequently misunderstood characterization of all lumber products. First, it does not represent a square foot (12 × 12 inches) of whatever piece of wood you pick up. It does represent a square foot of lumber at 1-inch thickness, again, going by nominal dimensions.

To calculate board feet use this formula: inches of width, times inches of depth, times feet of length, divided by twelve, equals board feet. For example, a 2 × 10, 12 feet long, would be calculated in board feet as follows: 2 (inches of width) × 10 (inches of depth) × 12 (feet of length) ÷ 12, or 240 ÷ 12 = 20 board feet.

Always confirm which measure is being used for the price quote—board feet or linear feet. On framing timbers that are 2 inches thick (nominal dimensions) the difference is substantial.

LUMBER DESIGN VALUES

The variation in timber sizes is great. But a piece of wood can be characterized in many ways other than length, width, and thickness, for instance, by weight, hardness, tightness of grain, percentage of defects, moisture content—the list of qualities is extensive.

However, the primary characteristics, the ones that directly affect the strength, durability, and safety of framing systems, are called design values. In essence, design values explain the natural, built-in characteristics of different kinds of wood. But instead of simply saying that pine is soft and pliable, while oak is hard and rigid, they provide precise, mathematical descriptions of the wood's mechanical properties.

These factors must be determined before you can use the detailed specifications of the span tables. There, you can establish precise limitations (a 2 × 10 over a 12-foot, 6-inch span, instead of a 2 × 8, for instance), determined by the natural characteristics (the design values) of the species and grade of wood you are using.

Don't be surprised, however, if a builder you hire, or a contractor you talk to is bypassing this information and working strictly from experience.

Many building inspectors work this

Framing Timber Dimensions

Nominal Size width-depth	Actual Size width-depth	Number of Linear Feet per Board Foot	Number of Board Feet per Single Length					
			8 feet	10 feet	12 feet	14 feet	16 feet	18 feet
2 × 2	1½ × 1½	3	2.7	3.3	4	4.7	5.3	6
2 × 3	1½ × 2½	2	4	5	6	7	8	9
2 × 4	1½ × 3½	1.5	5.3	6.7	8	9.3	10.7	12
2 × 6	1½ × 5½	1	8	10	12	14	16	18
2 × 8	1½ × 7½	.75	10.7	13.3	16	18.7	21.3	24
2 × 10	1½ × 9½	.60	13.3	16.7	20	23.3	26.7	30
2 × 12	1½ × 11½	.50	16	20	24	28	32	36
3 × 4	2½ × 3½	1	8	10	12	14	16	18
3 × 6	2½ × 5½	.67	12	15	18	21	24	27
3 × 8	2½ × 7½	.50	16	20	24	28	32	36
3 × 10	2½ × 9½	.40	20	25	30	35	40	45
3 × 12	2½ × 11½	.30	24	30	36	42	48	54
4 × 4	3½ × 3½	.75	10.7	13.3	16	18.7	21.3	24
4 × 6	3½ × 5½	.50	16	20	24	28	32	36
4 × 8	3½ × 7½	.38	21.3	26.7	32	37.3	42.7	48
4 × 10	3½ × 9½	.30	26.7	33.3	40	46.7	53.3	60
4 × 12	3½ × 11½	.25	32	40	48	56	64	72
6 × 6	5½ × 5½	.33	24	30	36	42	48	54
8 × 8	7½ × 7½	.19	42.7	53.3	64	74.7	85.3	96

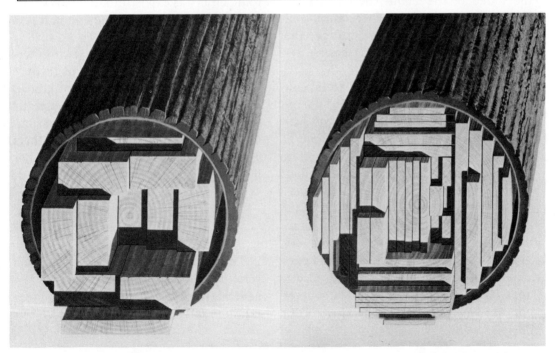

Figures 86 and 87. Different-size framing timbers for various uses are cut from different parts of the tree.

way too. They have seen enough framing and jumped up and down on enough new floors to know when a 2 × 8 will provide enough strength and when it will be too flexible. In fact, I've talked to several inspectors who have little or no knowledge of modulus of elasticity, and fiber stress in bending values, even though, as you'll see, these are only technical-sounding descriptions of very simple stresses. Yet these building inspectors would not undersize framing members because they select a 2 × 8 or a 2 × 10 the way you select second or third gear as you drive.

In a car there are exacting relationships between the tachometer's calibrations of engine revolutions per minute, and forward speed. Using this description, shifting gears may sound too technical to understand. Yet on many cars you shift by the sound of the engine, by experience. But when the behavior of structural materials is not second nature to you, the following tables are the only sensible resource for design decisions.

Don't be put off by the technical-sounding language they contain. Every field has its own jargon, and in construction, as in most professions, the jargon is only a convenient, shorthand description of things that are relatively easy to understand if you have a plain language explanation. Also, in the interest of keeping potentially confusing statistics to a minimum, only commonly used species and grades of wood are listed here. If the particular species and grade you plan to use isn't included, consult the list of sources for further information on page 220.

Lumber associations can provide voluminous technical material.

To review the basics, there are many kinds of wood. Douglas fir, hemlock, spruce, and redwood are some of the most common framing materials. There are others. And each species has its own special characteristics. Their wood fibers behave uniquely under stress, and some species are simply stronger than others.

In addition, each species is available in many different grades, from selects, generally called appearance grades, which are suitable for trim, paneling, and siding, to light framing, utility grades, which are used in the structural skeleton and buried under finishing materials. Grade differentiation, for example, from one Douglas fir 2 × 6 to another is influenced by such factors as where the tree grew, the environment it was exposed to, the part of the tree it was cut from, and how long it was seasoned.

These variations can make two different grades of the same species as distinct as two different species that are graded equally. It can make some Douglas fir 2 × 6's suitable for exposed, dimensionally stable, fixed-glass frames, and others suitable only for concealed blocking or light-duty concrete form supports.

INTERPRETING A LUMBER DESIGN VALUE TABLE

The table on page 151 gives data about selected species of wood. It serves two general purposes. First, it indicates properties of different species that determine how the lumber will behave, and therefore, how it should be used. Second, two particular values—(the fiber stress in bending (F_b), and modulus of elasticity (E) values—serve as the key elements for determining span limits, a principal consideration of framing design.

For comparative purposes all figures are for lumber graded construction, which is a commonly available framing grade. Here are the characteristics covered by the table.

Extreme Fiber Stress in Bending (F_b)

When a beam supported at each end is loaded from above, it tends to sag in the middle. Wood fibers along the bottom edge of the beam will stretch as the load from above increases, and the beam bows downward. This stretching tension is indicated in pounds per square inch (psi) for the fibers farthest from the load.

SINGLE F_b These figures are values for individual timbers (a single post or girder, for instance) that will carry its full design load independently.

REPETITIVE F_b These figures are also values for individual framing members (joists, rafters, studs, etc.), but only when they are used in combination with similar framing timbers (at least three of them), spaced not more than 24 inches apart, and joined by a load-distributing material like plywood subflooring or sheathing. This integration of building components provides a 15 percent increase in load-carrying capacity, or, conversely, a 15 percent decrease in the strength requirements of any one member in the group. Other strength-increasing factors above those specified in the tables are: 15 percent for a two-month load duration, as for snow; 25 percent for a seven-day duration, as for severe loading of construction materials; $33\frac{1}{3}$ percent for severe wind load or earthquake; and 100 percent for impact resistance.

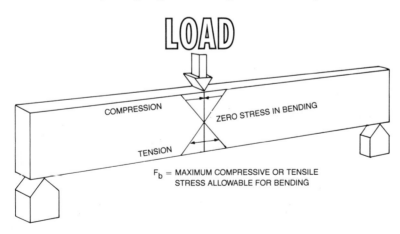

Figure 88. Extreme fiber stress in bending (F_b)

Modulus of Elasticity (E)

Technically, this characteristic, which also varies with wood species and grade, is the ratio of stress (or force per unit area) to strain (or deformation per unit length), expressed in psi. In plain language, it is the material's resistance to downward bending (called deflection) when loaded from above. Span figures for joists and rafters are often noted as limited by deflection, i.e., the maximum unsupported span that can be bridged without exceeding a deflection limit of $1/360$ of the span length. This translates to roughly 1 inch of deflection in a 30-foot span, $5/8$ inch deflection over 20 feet, and $5/16$ inch deflection over 10 feet, all acceptable amounts.

Horizontal Shear (F_v)

This is a horizontal stress running along the length of internal fibers in a beam. The effect is similar to the way playing cards slide over each other when the deck is bent. The wood's resistance to horizontal shear, in psi, makes it bend as one unit.

Compression Perpendicular to Grain ($f_c\perp$)

This stress measures the crushing of wood fibers in a beam. For example, if a horizontal 4 × 10 joist supports a vertical 4 × 4 post that is carrying a heavy load from the roof, the concentrated weight can compress the uppermost fibers in the joist where the 4 × 4 presses down onto it. This usually marginal effect can be significant when vertical loads are carried through successive layers of horizontal timbers (girders, floor joists, etc.), instead of through a single, vertical post.

Compression Parallel to Grain (F_c)

Parallel compression stresses the grain longitudinally. For example, the vertical fibers in a 4 × 4 post, placed between the foundation sill and the second floor joists, will tend to compress from the load it carries. This effect (marginal) is equal across the full cross section of the timber.

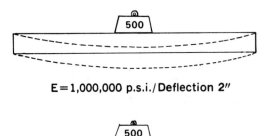

E = 1,000,000 p.s.i. / Deflection 2″

E = 2,000,000 p.s.i. / Deflection 1″

Figure 89. Modulus of elasticity (E)

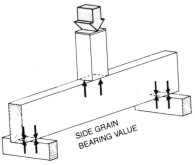

Figure 90. Compression perpendicular to grain ($F_c\perp$)

Figure 91. Compression parallel to grain (F_c)

Design Values and Properties of Framing Lumber

Wood and Species (Construction grade)	Extreme Fiber Stress in Bending (psi) Single/Rep.	Horizontal Shear (psi)	Compression Perpendicular to Grain (psi)	Compression Parallel to Grain (psi)	Modulus of Elasticity (psi)	Weight When Green pounds/cubic foot	Weight at 19 Percent Moisture Content (lbs./cu. ft.)	Moisture Content When Green percentage	Shrinkage from Green to 20 Percent Moisture Content percentage
Western Cedar	775–875	75	265	850	900,000	26.4	24	37	0.8
Douglas Fir-Larch	1,050–1,200	95	385	1,150	1,500,000	38.2	35	38	1.7
Douglas Fir, South	1,000–1,150	90	335	1,000	1,100,000	36.3	32.4	48	1.4
Hem-Fir combined	825–975	75	245	925	1,200,000	36.8	31.1	61	1.2
Eastern Hemlock	900–1,050	85	360	950	1,000,000	43.4	29.7	111	1.0
Western Hemlock	925–1,050	90	280	1,050	1,300,000	37.2	30.1	74	1.4
Eastern White Pine	700–800	70	220	750	1,000,000	35.1	26	73	0.8
Southern Pine	1,000–1,150	100	405	1,100	1,400,000	50.2	42.3	63	1.6
California Redwood	825–950	80	270	925	900,000	45.6	29.3	112	0.9
Eastern Spruce	775–875	70	265	775	1,200,000	33	29.8	50	1.1
Sitka Spruce	800–925	75	280	825	1,200,000	32	28.7	42	1.4

Average Weight When Green

Green weight represents the combined weight of wood and the water it contains before natural seasoning or kiln-drying. Green wood contains water in the individual cell walls, and in the cavities between fibers, which dry first. At 25 to 30 percent moisture content, shrinkage will start as the cavities are emptied, and water is pulled from the cell walls.

Average Weight at 19 Percent Moisture Content

These figures can be used, as averages, to calculate dead loads, or the weights of structural components like joists and rafters, as though they were weighed on a scale. It pays to spot-check dimensional timbers in the lumberyard. Studs that are obviously heavier than normal are likely waterlogged, carrying moisture in excess of the standard, accepted 19 percent level for framing members, and should be left behind. On lumber grade stamps showing the grading association, mill number, grade, and species, there are three marks that indicate different moisture levels: S-GRN, for moisture content over 19 percent, S-DRY, for moisture content not exceeding 19 percent, and MC 15, for particularly stable wood with a moisture content below 15 percent.

Moisture Content When Green

This percentage figure is arrived at by comparing water weight to the weight of the wood only, without any water. It is possible, therefore, to have a moisture content of more than 100 percent, where water represents more of the total weight (water and wood), than the dry wood.

Shrinkage from Green to 20 Percent Moisture Content

These figures measure radial shrinkage as the wood dries from its natural state. Longitudinal shrinkage (along the length of the timber), is negligible, averaging about .002 percent, or less than ¼ inch over 12 feet.

Remember, even if you are familiar with construction techniques you should get professional guidance with particularly unique designs or special environmental conditions. In any case, even the most typical framing design should be checked by the building inspector before the plans are approved.

Two crucial categories on the table limit frame design more severely and more frequently than weight, moisture content, or other wood properties. They are fiber stress in bending—repetitive, and modulus of elasticity. These values must be known before you can use the span tables (pages 156–64).

These tables, which govern the design of joists and rafters, can be used in several ways. If a number of wood species and grades are locally available at competitive prices, you can check through the span tables for any combination of values (spans limited by bending, F_b and by deflection, E) that permit the design you want. After these design values are determined, it is a simple matter to use the design value table to pick one of the species

Properties and Design Values of Framing Members

and grades that meets or surpasses these requirements.

Conversely, when species and grade selection are limited, the F_b and E values derived from the design value table will also be limited. When you check these values on the span limit tables you may have to alter the design from, for instance, a 24-foot living room to a 22-foot room.

The most common situation, however, provides only one option. You call several lumberyards, and they all have one species in one grade, take it or leave it. In this case you look up the appropriate F_b and E values, and proceed to the span tables to find out what sizes of timbers you will need to bridge the variety of spans called for in the design.

Wood is a flexible building material. This is reflected in the endless combinations of species, grades, sizes, and spans in the tables. This flexibility also provides an arena in which you must make choices and do the balancing act between cost and appearance, between efficiency and strength, between expediency and durability.

Also remember that F_b (bending) and E (deflection) values are separate and distinct. There is no constant mathematical connection between them. One species and grade of wood may have an E value permitting a 12-foot span for a 2 × 8 floor joist, but an F_b value permitting only an 11-foot, 6-inch span. With a different species and grade, the converse may be true.

All wood species and grades are different. To be safe you must check both F_b and E values and determine (on the span table) which is the most limiting factor.

INTERPRETING SPAN TABLES

Standardized span tables are an essential tool of frame design. They contain accepted figures from the National Design Specification at the National Forest Products Association (see page 220). The tables are completely straightforward. Given specific F_b and E values, given the lumber size (2 × 4, 2 × 10, and so forth), and given the framing module (16 or 24 inches center to center), the tables show exactly how long a timber can be. But it's not quite that simple because there is no single, all-encompassing table.

Several span tables are necessary because the limitations on a 2 × 6, for instance, are quite different from one situation to another. Used as a floor joist in a high-traffic, furniture-loaded living area, it will be subject to far more stress than it would be if installed as a high slope rafter.

The span tables make several of these distinctions: floor joists in living areas; ceiling joists with and without attic storage space (if you plan, at some point, to convert this dead space to living space you should use figures for floor joists in living areas); flat or low slope rafters with 20-, 30-, and 40-pound loads (depending on ceiling and roofing materials used as well as environmental conditions); and medium or high slope rafters with the same load distinctions.

Additionally, there are several factors that can be used to modify span limits in particular situations. These are really proven and accepted extensions of experience and common sense expressed in precise terms.

DEFLECTION The standard factor for the high level of stiffness required on floor and ceiling joists is L/360, or length of span in inches, divided by 360. This means that the timber should not bow downward more than 1/360 of its length. However, this factor may be modified to L/240 in situations where stiffness is considered less crucial, or where the load is light. A further reduction to L/180 may be permitted for high slope rafters where no ceiling load is carried, and where live loads of workmen and building materials are carried only for short durations. Always consult local building codes for requirements. And if you are buying a new house or having an addition built, you will certainly get the attention of the builder by asking him to tell you which joists and rafters are spans limited by L/360 or L/240.

BENDING Single-use fiber stress in bending values are only for situations where a beam or girder carries its full design load independently. Repetitive-use bending values for multiple applications in a framing system are increased by 15 percent due to the extra strength derived from multiple use.

DURATION OF LOAD The requirements given for normal load duration may be modified as follows: 15 percent increase for two months (snow); 25 percent for seven days (material loading during construction); $33\frac{1}{3}$ percent for severe wind or earthquake; and 100 percent for impact. Remember, when a structure is subject to special environmental conditions, extreme care should be used in frame design and expert help sought in order to determine span limits and other lumber characteristics that will make the building completely safe.

In areas of California particularly prone to flash fires, key structural beams are sometimes made of heartwood, which is the densest part of the tree. It is likely to smolder slowly and retain its structural integrity longer than a conventional beam. It is more costly but offers a special kind of protection against unique environmental conditions.

CONVERSION FOR TRUE RAFTER SIZE

In Figure 92 you will find an easy-to-use diagram for converting span limit sizes to true rafter sizes. On the tables, span limits for both high and low slope rafters are computed as horizontal distances. Obviously, a sloping rafter will be longer than this dimension just the way the hypotenuse of a triangle is longer than the base leg. The Rafter Conversion Diagram is used to convert the horizontal rafter measure given in the span tables to sloping distance, or the actual rafter size, so you will know what size timbers to order.

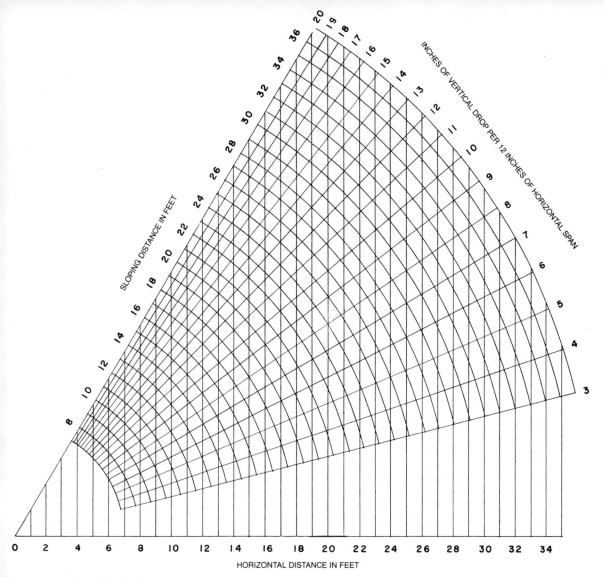

Figure 92. Rafter conversion diagram
This diagram shows the relationship of three factors: sloping distance (the true rafter size), horizontal distance (the length between roof ridge and side wall), and rise (expressed in the inches of vertical drop per 12 inches of horizontal span). Any one value can be found if you know the other two. Typically, a side elevation or framing plan on the blueprints will show the rise and span dimensions from which you can determine rafter size. For example, the horizontal distance from the center of the roof to the sidewall is 20 feet. The roof slopes down 8 inches for every 12 inches of horizontal distance. To find true rafter size locate the intersection of 20 (horizontal distance) and 8 (slope in 12 inches), and follow the intersecting arc to find the sloping distance.

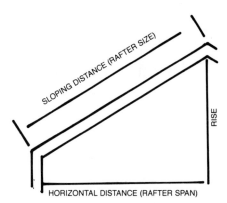

Floor Joist Spans: Sleeping Areas, Attic Floors

30 pounds per square foot live load, limited by deflection, 1/360 of span

NOTE: required bending values appear after span limit, which is shown in feet-inches.

Joist Size (in.)	Joist Spacing (in.)	Modulus of Elasticity (E) Design Values (in psi)						
		900,000	1,000,000	1,100,000	1,200,000	1,300,000	1,400,000	1,500,000
2 × 6	16	8-10/830	9-2/890	9-6/950	9-9/1000	10-0/1060	10-3/1110	10-6/1160
2 × 6	24	7-9/950	8-0/1020	8-3/1080	8-6/1150	8-9/1210	8-11/1270	9-2/1330
2 × 8	16	11-8/830	12-1/890	12-6/950	12-10/1000	13-2/1060	13-6/1110	13-10/1160
2 × 8	24	10-2/950	10-7/1020	10-11/1080	11-3/1150	11-6/1210	11-10/1270	12-1/1330
2 × 10	16	14-11/770	15-5/890	15-11/950	16-5/1000	16-10/1060	17-3/1110	17-8/1160
2 × 10	24	13-0/950	13-6/1020	13-11/1080	14-4/1150	14-8/1210	15-1/1270	15-5/1330
2 × 12	16	18-1/830	18-9/890	19-4/950	19-11/1000	20-6/1060	21-0/1110	21-6/1160
2 × 12	24	15-10/950	16-5/1020	16-11/1080	17-5/1150	17-11/1210	18-4/1210	18-9/1330

Floor Joist Spans: Living Areas

40 pounds per square foot live load, limited by deflection, 1/360 of span

NOTE: required bending values appear after span limit, which is shown in feet-inches.

Joist Size (inches)	Joist Spacing (inches)	Modulus of Elasticity (E) Design Values (in psi)						
		900,000	1,000,000	1,100,000	1,200,000	1,300,000	1,400,000	1,500,000
2 × 6	16	8-0/860	8-4/920	8-7/980	8-10/1040	9-1/1090	9-4/1150	9-6/1200
2 × 6	24	7-0/980	7-3/1050	7-6/1120	7-9/1190	7-11/1250	8-2/1310	8-4/1380
2 × 8	16	10-7/850	11-0/920	11-4/980	11-8/1040	12-0/1090	12-3/1150	12-7/1200
2 × 8	24	9-3/980	9-7/1050	9-11/1120	10-2/1190	10-6/1250	10-9/1310	11-0/1380
2 × 10	16	13-6/850	14-0/920	14-6/980	14-11/1040	15-3/1090	15-8/1150	16-0/1200
2 × 10	24	11-10/980	12-3/1050	12-8/1120	13-0/1190	13-4/1250	13-8/1310	14-0/1380
2 × 12	16	16-5/860	17-0/920	17-7/980	18-1/1040	18-7/1090	19-1/1150	19-6/1200
2 × 12	24	14-4/980	14-11/1050	15-4/1120	15-10/1190	16-3/1250	16-8/1310	17-0/1380

Ceiling Joist Spans: No Attic Storage

10 pounds per square foot live load, limited by deflection, 1/360 of span

NOTE: required bending values appear after span limit, which is shown in feet-inches. Use top figures for drywall ceilings, bottom figures for plaster ceilings.

Joist Size (inches)	Joist Spacing (inches)	Modulus of Elasticity (E) Design Values (in psi)						
		900,000	1,000,000	1,100,000	1,200,000	1,300,000	1,400,000	1,500,000
2 × 4	16	9-4/850 8-1/650	9-8/910 8-5/690	9-11/970 8-8/740	10-3/1030 8-11/780	10-6/1080 9-2/830	10-9/1140 9-5/870	11-0/1190 9-8/910
2 × 4	24	8-1/970 7-1/740	8-5/1040 7-4/790	8-8/1110 7-7/850	8-11/1170 7-10/900	9-2/1240 8-0/950	9-5/1300 8-3/990	9-8/1360 8-5/1040
2 × 6	16	14-7/850 12-9/650	15-2/910 13-3/690	15-7/970 13-8/740	16-1/1030 14-1/780	16-6/1080 14-5/830	16-11/1140 14-9/870	17-4/1190 15-2/910
2 × 6	24	12-9/970 11-2/740	13-3/1040 11-7/790	13-8/1110 11-11/850	14-1/1170 12-3/900	14-5/1240 12-7/950	14-9/1300 12-11/990	15-2/1360 13-3/1040
2 × 8	16	19-3/850 16-10/650	19-11/910 17-5/690	20-7/970 18-0/740	21-2/1030 18-6/780	21-9/1080 19-0/830	22-4/1140 19-6/870	22-10/1190 19-11/910
2 × 8	24	16-10/970 14-8/740	17-5/1040 15-3/790	18-0/1110 15-9/850	18-6/1170 16-2/900	19-0/1240 16-7/950	19-6/1300 17-0/990	19-11/1360 17-5/1040
2 × 10	16	24-7/850 21-6/650	25-5/910 22-3/690	26-3/970 22-11/740	27-10/1030 23-8/780	27-9/1080 24-3/830	28-6/1140 24-10/870	29-1/1190 25-5/910
2 × 10	24	21-6/970 18-9/740	22-3/1040 19-5/790	22-11/1110 20-1/850	23-8/1170 20-8/900	24-3/1240 21-2/950	24-10/1300 21-9/990	25-5/1360 22-3/1040

Ceiling Joist Spans: Limited Attic Storage
20 pounds per square foot live load, limited by deflection, 1/360 of span

NOTE: required bending values appear after span limit, which is shown in feet-inches. Use top figures for drywall ceilings, bottom figures for plaster ceilings.

Joist Size (inches)	Joist Spacing (inches)	\multicolumn{7}{c}{Modulus of Elasticity (E) Design Values (in psi)}
		900,000	1,000,000	1,100,000	1,200,000	1,300,000	1,400,000	1,500,000
2 × 4	16	7-5/1070	7-8/1140	7-11/1220	8-1/1290	8-4/1360	8-7/1430	8-9/1500
		6-5/810	6-8/870	6-11/930	7-1/990	7-3/1040	7-6/1090	7-8/1140
2 × 4	24	6-5/1220	6-8/1310	6-11/1400	7-1/1480	7-3/1560	7-6/1640	7-8/1720
		5-8/930	1-10/1000	6-0/1070	6-2/1130	6-4/1190	6-6/1250	6-8/1310
2 × 6	16	11-7/1070	12-0/1140	12-5/1220	12-9/1290	13-1/1360	13-5/1430	13-9/1500
		10-2/810	10-6/870	10-10/930	11-2/990	11-5/1040	11-9/1090	12-0/1140
2 × 6	24	10-2/1220	10-6/1310	10-10/1400	11-2/1480	11-5/1560	11-9/1640	12-0/1720
		8-10/930	9-2/1000	9-6/1070	9-9/1130	10-0/1190	10-3/1250	10-6/1310
2 × 8	16	15-3/1070	15-10/1140	16-4/1220	16-10/1290	17-3/1360	17-9/1430	18-2/1500
		13-4/810	13-10/870	14-3/930	14-8/990	15-1/1040	15-6/1090	15-10/1140
2 × 8	24	13-4/1220	13-10/1310	14-3/1400	14-8/1480	15-1/1560	15-6/1640	15-10/1720
		11-8/930	12-1/1000	12-6/1070	12-10/1130	13-2/1190	13-6/1250	13-10/1310
2 × 10	16	19-6/1070	20-2/1140	20-10/1220	21-6/1290	22-1/1360	22-7/1430	23-2/1500
		17-0/810	17-8/870	18-3/930	18-9/990	19-3/1040	19-9/1090	20-2/1140
2 × 10	24	17-0/1220	17-8/1310	18-3/1400	18-9/1480	19-3/1560	19-9/1640	20-2/1720
		14-11/930	15-5/1000	15-11/1070	16-5/1130	16-10/1190	17-3/1250	17-8/1310

Rafter Spans: Flat or Low Slope
20 pounds per square foot live load, limited by bending

NOTE: required deflection values (1/240 of span shown in millions) appear after span limit, which is shown in feet-inches. Use top figures for drywall ceilings, bottom figures for no ceiling load.

Rafter Size (inches)	Rafter Spacing (inches)	Extreme Fiber Stress in Bending Design Values (in psi)						
		800	900	1,000	1,100	1,200	1,300	1,400
2 × 6	16	9-4/0.46 10-0/0.58	9-10/0.55 10-8/0.70	10-5/0.65 11-3/0.82	10-11/0.75 11-9/0.94	11-5/0.85 12-4/1.07	11-10/0.96 12-10/1.21	12-4/1.07 13-3/1.35
2 × 6	24	7-7/0.38 8-2/0.48	8-1/0.45 8-8/0.57	8-6/0.53 9-2/0.67	8-11/0.61 9-7/0.77	9-4/0.70 10-0/0.88	9-8/0.78 10-5/0.99	10-0/0.88 10-10/1.10
2 × 8	16	12-3/0.46 13-3/0.58	13-0/0.55 14-0/0.70	13-8/0.65 14-10/0.82	14-4/0.75 15-6/0.94	15-0/0.85 16-3/1.07	15-7/0.96 16-10/1.21	16-3/1.07 17-6/1.35
2 × 8	24	10-0/0.38 10-10/0.48	10-7/0.45 11-6/0.57	11-2/0.53 12-1/0.67	11-9/0.61 12-8/0.77	12-3/0.70 13-3/0.88	12-9/0.78 13-9/0.99	13-3/0.88 14-4/1.10
2 × 10	16	15-8/0.46 16-11/0.58	16-7/0.55 17-11/0.70	17-6/0.65 18-11/0.82	18-4/0.75 19-10/0.94	19-2/0.85 20-8/1.07	19-11/0.96 21-6/1.21	20-8/1.07 22-4/1.35
2 × 10	24	12-9/0.38 13-9/0.48	13-6/0.45 14-8/0.57	14-3/0.53 15-5/0.67	15-0/0.61 16-2/0.77	15-8/0.70 16-11/0.88	16-3/0.78 17-7/0.99	16-11/0.88 18-3/1.10
2 × 12	16	19-0/0.46 20-6/0.58	20-1/0.55 21-9/0.70	21-3/0.65 23-0/0.82	22-4/0.75 24-1/0.94	23-3/0.85 25-2/1.07	24-3/0.96 26-1/1.21	25-2/1.07 27-2/1.35
2 × 12	24	15-6/0.38 16-9/0.48	16-6/0.45 17-9/0.57	17-4/0.53 18-9/0.67	18-2/0.61 19-8/0.77	19-0/0.70 20-6/0.88	19-10/0.78 21-5/0.99	20-6/0.88 22-2/1.10

Rafter Spans: Flat or Low Slope
30 pounds per square foot live load, limited by bending

NOTE: required deflection values (1/240 of span shown in millions) appear after span limit, which is shown in feet-inches. Use top figures for drywall ceilings, bottom figures for no ceiling load.

Rafter Size (inches)	Rafter Spacing (inches)	Extreme Fiber Stress in Bending Design Values (in psi)							
		800	900	1,000	1,100	1,200	1,300	1,400
2 × 6	16	8-2/0.48 8-8/0.57	8-8/0.57 9-3/0.68	9-2/0.67 9-9/0.80	9-7/0.77 10-2/0.92	10-0/0.88 10-8/1.05	10-5/0.99 11-1/1.18	10-10/1.10 11-6/1.18
2 × 6	24	6-8/0.39 7-1/0.46	7-1/0.46 7-6/0.55	7-6/0.54 7-11/0.65	7-10/0.63 8-4/0.75	8-2/0.72 8-8/0.85	8-6/0.81 9-1/0.96	8-10/0.90 9-5/1.08
2 × 8	16	10-10/0.48 11-6/0.57	11-6/0.57 12-2/0.68	12-1/0.67 12-10/0.80	12-8/0.77 13-5/0.92	13-3/0.88 14-0/1.05	13-9/0.99 14-7/1.18	14-4/1.10 15-2/1.32
2 × 8	24	8-10/0.39 9-4/0.46	9-4/0.46 9-11/0.55	9-10/0.54 10-6/0.65	10-4/0.63 11-0/0.75	10-10/0.72 11-6/0.85	11-3/0.81 11-11/0.96	11-8/0.90 12-5/1.08
2 × 10	16	13-9/0.48 14-8/0.57	14-8/0.57 15-6/0.68	15-5/0.67 16-4/0.80	16-2/0.77 17-2/0.92	16-11/0.88 17-11/1.05	17-7/0.99 18-8/1.18	18-3/1.10 19-4/1.32
2 × 10	24	11-3/0.39 11-11/0.46	11-11/0.46 12-8/0.55	12-7/0.54 13-4/0.65	13-2/0.63 14-0/0.75	13-9/0.72 14-8/0.85	14-4/0.81 15-3/0.96	14-11/0.90 15-10/1.08
2 × 12	16	16-9/0.48 17-9/0.57	17-9/0.57 18-10/0.68	18-9/0.67 19-11/0.80	19-8/0.77 20-10/0.92	20-6/0.88 21-9/1.05	21-5/0.99 22-8/1.18	22-2/1.10 23-6/1.32
2 × 12	24	13-8/0.39 14-6/0.46	14-6/0.46 15-5/0.55	15-4/0.54 16-3/0.65	16-1/0.63 17-0/0.75	16-9/0.72 17-9/0.85	17-5/0.81 18-6/0.96	18-1/0.90 19-3/1.08

Rafter Spans: Flat or Low Slope

40 pounds per square foot live load, limited by bending

NOTE: required deflection values (1/240 of span shown in millions) appear after span limit, which is shown in feet-inches. Use top figures for drywall ceilings, bottom figures for no ceiling load.

Rafter Size (inches)	Rafter Spacing (inches)	Extreme Fiber Stress in Bending (in psi)						
		800	900	1,000	1,100	1,200	1,300	1,400
2 × 6	16	7-5/0.47 7-9/0.54	7-10/0.56 8-3/0.65	8-3/0.66 8-8/0.76	8-8/0.76 9-1/0.88	9-1/0.86 9-6/1.00	9-5/0.98 9-11/1.12	9-10/1.09 10-3/1.26
2 × 6	24	6-1/0.38 6-4/0.44	6-5/0.46 6-9/0.53	6-9/0.54 7-1/0.62	7-1/0.62 7-5/0.71	7-5/0.71 7-9/0.81	7-9/0.80 8-1/0.92	8-0/0.89 8-5/1.03
2 × 8	16	9-9/0.47 10-3/0.54	10-4/0.56 10-11/0.65	10-11/0.66 11-6/0.76	11-6/0.76 12-0/0.88	12-0/0.86 12-7/1.00	12-6/0.98 13-1/1.12	12-11/1.09 13-7/1.26
2 × 8	24	8-0/0.38 8-4/0.44	8-6/0.46 8-11/0.53	8-11/0.54 9-4/0.62	9-4/0.62 9-10/0.71	9-9/0.71 10-3/0.81	10-2/0.80 10-8/0.92	10-7/0.89 11-1/1.03
2 × 10	16	12-6/0.47 13-1/0.54	13-3/0.56 13-11/0.65	13-11/0.66 14-8/0.76	14-8/0.76 15-4/0.88	15-3/0.86 16-0/1.00	15-11/0.98 16-8/1.12	16-6/1.09 17-4/1.26
2 × 10	24	10-2/0.38 10-8/0.44	10-10/0.46 11-4/0.53	11-5/0.54 11-11/0.62	11-11/0.62 12-6/0.71	12-6/0.71 13-1/0.81	13-0/0.80 13-7/0.92	13-6/0.89 14-2/1.03
2 × 12	16	15-2/0.47 15-11/0.54	16-1/0.56 16-11/0.65	17-0/0.66 17-9/0.76	17-9/0.76 18-8/0.88	18-7/0.86 19-6/1.00	19-4/0.98 20-3/1.12	20-1/1.09 21-1/1.26
2 × 12	24	12-5/0.38 13-0/0.44	13-2/0.46 13-9/0.53	13-10/0.54 14-6/0.62	14-6/0.62 15-3/0.71	15-2/0.71 15-11/0.81	15-9/0.80 16-7/0.92	16-5/0.89 17-2/1.03

Rafter Spans: Medium or High Slope (Over 3 in 12), No Ceiling Load
20 pounds per square foot live load, limited by bending

Rafter Size (inches)	Rafter Spacing (inches)	Extreme Fiber Stress in Bending (in psi)						
		800	900	1,000	1,100	1,200	1,300	1,400
2 × 4	16	6-9/0.51 5-11/0.35	7-2/0.61 6-3/0.41	7-6/0.72 6-7/0.49	7-11/0.83 6-11/0.56	8-3/0.94 7-3/0.64	8-7/1.06 7-6/0.72	8-11/1.19 7-10/0.80
2 × 4	24	5-6/0.42 4-10/0.28	5-10/0.50 5-1/0.34	6-2/0.59 5-5/0.40	6-5/0.68 5-8/0.46	6-9/0.77 5-11/0.52	7-0/0.87 6-2/0.59	7-3/0.97 6-5/0.66
2 × 6	16	10-7/0.51 9-4/0.35	11-3/0.61 9-10/0.41	11-10/0.72 10-5/0.49	12-5/0.83 10-11/0.56	13-0/0.94 11-5/0.64	13-6/1.06 11-10/0.72	14-0/1.19 12-4/0.80
2 × 6	24	8-8/0.42 7-7/0.28	9-2/0.50 8-1/0.34	9-8/0.59 8-6/0.40	10-2/0.68 8-11/0.46	10-7/0.77 9-4/0.52	11-0/0.87 9-8/0.59	11-5/0.97 10-0/0.66
2 × 8	16	13-11/0.51 12-3/0.35	14-10/0.61 13-0/0.41	15-7/0.72 13-8/0.49	16-4/0.83 14-4/0.56	17-1/0.94 15-0/0.64	17-9/1.06 15-7/0.72	18-5/1.19 16-3/0.80
2 × 8	24	11-5/0.42 10-0/0.28	12-1/0.50 10-7/0.34	12-9/0.59 11-2/0.40	13-4/0.68 11-9/0.46	13-11/0.77 12-3/0.52	14-6/0.87 12-9/0.59	15-1/0.97 13-3/0.66
2 × 10	16	17-10/0.51 15-8/0.35	18-11/0.61 16-7/0.41	19-11/0.72 17-6/0.49	20-10/0.83 18-4/0.56	21-10/0.94 19-2/0.64	22-8/1.06 19-11/0.72	23-7/1.19 20-8/0.80
2 × 10	24	14-6/0.42 12-9/0.28	15-5/0.50 13-6/0.34	16-3/0.59 14-3/0.40	17-1/0.68 15-0/0.46	17-10/0.77 15-8/0.52	18-6/0.87 16-3/0.59	19-3/0.97 16-11/0.66

NOTE: required deflection values (1/180 of span shown in millions) appear after span limit, which is shown in feet-inches. Use top figures for light roof covering (7 pounds psf dead load), bottom figures for heavy roofing (15 pounds psf dead load).

Rafter Spans: Medium or High Slope (over 3 in 12), No Ceiling Load
30 pounds per square foot live load, limited by bending

NOTE: required deflection values (1/180 of span shown in millions) appear after span limit, which is shown in feet-inches. Use top figures for light roof covering (7 pounds psf dead load), bottom figures for heavy roofing (15 pounds psf dead load).

Rafter Size (inches)	Rafter Spacing (inches)	Extreme Fiber Stress in Bending (in psi)						
		800	900	1,000	1,100	1,200	1,300	1,400
2 × 4	16	5-9/0.48 5-3/0.36	6-1/0.57 5-6/0.43	6-5/0.67 5-10/0.50	6-9/0.77 6-1/0.58	7-1/0.88 6-5/0.66	7-4/0.99 6-8/0.74	7-7/1.11 6-11/0.83
2 × 4	24	4-8/0.39 4-3/0.29	5-0/0.47 4-6/0.35	5-3/0.55 4-9/0.41	5-6/0.63 5-0/0.47	5-9/0.72 5-3/0.54	6-0/0.81 5-5/0.61	6-3/0.91 5-8/0.68
2 × 6	16	9-1/0.48 8-2/0.36	9-7/0.57 8-8/0.43	10-1/0.67 9-2/0.50	10-7/0.77 9-7/0.58	11-1/0.88 10-0/0.66	11-6/0.99 10-5/0.74	12-0/1.11 10-10/0.83
2 × 6	24	7-5/0.39 6-8/0.29	7-10/0.47 7-1/0.35	8-3/0.55 7-6/0.41	8-8/0.63 7-10/0.47	9-1/0.72 8-2/0.54	9-5/0.81 8-6/0.61	9-9/0.91 8-10/0.68
2 × 8	16	11-11/0.48 10-10/0.36	12-8/0.57 11-6/0.43	13-4/0.67 12-1/0.50	14-0/0.77 12-8/0.58	14-7/0.88 13-3/0.66	15-2/0.99 13-9/0.74	15-9/1.11 14-4/0.83
2 × 8	24	9-9/0.39 8-10/0.29	10-4/0.47 9-4/0.35	10-11/0.55 9-10/0.41	11-5/0.63 10-4/0.47	11-11/0.72 10-10/0.54	12-5/0.81 11-3/0.61	12-10/0.91 11-8/0.68
2 × 10	16	15-2/0.53 13-9/0.36	16-2/0.63 14-8/0.43	17-0/0.74 15-5/0.50	17-10/0.85 16-2/0.58	18-7/0.97 16-11/0.66	19-5/1.09 17-7/0.74	20-1/1.22 18-3/0.83
2 × 10	24	12-5/0.39 11-3/0.29	13-2/0.47 11-11/0.35	13-11/0.55 12-7/0.41	14-7/0.63 13-2/0.47	15-2/0.72 13-9/0.54	15-10/0.81 14-4/0.61	16-5/0.91 14-11/0.68

Rafter Spans: Medium or High Slope (over 3 in 12), No Ceiling Load
40 pounds per square foot live load, limited by bending

NOTE: required deflection values (1/180 of span shown in millions) appear after span limit, which is shown in feet-inches. Use top figures for light roof covering (7 pounds psf dead load), bottom figures for heavy roof covering (15 pounds psf dead load).

Rafter Size (inches)	Rafter Spacing (inches)	Extreme Fiber Stress in Bending (in psi)						
		800	900	1,000	1,100	1,200	1,300	1,400
2 × 4	16	5-1/0.45 4-9/0.35	5-5/0.53 5-0/0.42	5-8/0.62 5-3/0.49	6-0/0.72 5-6/0.57	6-3/0.82 5-9/0.65	6-6/0.93 6-0/0.73	6-9/1.03 6-3/0.82
2 × 4	24	4-2/0.36 3-10/0.29	4-5/0.44 4-1/0.34	4-8/0.51 4-4/0.40	4-11/0.59 4-6/0.46	5-1/0.67 4-9/0.53	5-4/0.76 4-11/0.60	5-6/0.84 5-1/0.67
2 × 6	16	8-0/0.45 7-5/0.35	8-6/0.53 7-10/0.42	9-0/0.62 8-3/0.49	9-5/0.72 8-8/0.57	9-10/0.82 9-1/0.65	10-3/0.93 9-5/0.73	10-7/1.03 9-10/0.82
2 × 6	24	6-7/0.36 6-1/0.29	6-11/0.44 6-5/0.34	7-4/0.51 6-9/0.40	7-8/0.59 7-1/0.46	8-0/0.67 7-5/0.53	8-4/0.76 7-9/0.60	8-8/0.84 8-0/0.67
2 × 8	16	10-7/0.45 9-9/0.35	11-3/0.53 10-4/0.42	11-10/0.62 10-11/0.49	12-5/0.72 11-6/0.57	12-11/0.82 12-0/0.65	13-6/0.93 12-6/0.73	14-0/1.03 12-11/0.82
2 × 8	24	8-8/0.36 8-0/0.29	9-2/0.44 8-6/0.34	9-8/0.51 8-11/0.40	10-2/0.59 9-4/0.46	10-7/0.67 9-9/0.53	11-0/0.76 10-2/0.60	11-5/0.84 10-7/0.67
2 × 10	16	13-6/0.45 12-6/0.35	14-4/0.53 13-3/0.42	15-1/0.62 13-11/0.49	15-10/0.72 14-8/0.57	16-6/0.82 15-3/0.65	17-2/0.93 15-11/0.73	17-10/1.03 16-6/0.82
2 × 10	24	11-0/0.36 10-2/0.29	11-8/0.44 10-10/0.34	12-4/0.51 11-5/0.40	12-11/0.59 11-11/0.46	13-6/0.67 12-6/0.53	14-1/0.76 13-0/0.60	14-7/0.84 13-6/0.67

DESIGN GUIDELINES

In most localities either a licensed architect's or engineer's seal is required on construction plans before a building permit is issued. There are exceptions. Many building departments, particularly those in rural areas with minimal zoning and code restrictions, will issue permits on unorthodox plans. In other words, if you neatly sketch out the framing plan on a large sheet of paper, and notate the proper, code-approved timber sizes, the homemade blueprint is likely to be sufficient.

It is also possible, again only in some areas, for the requirement of stamped blueprints to be lifted when the total cost of the job is less than a specific dollar amount, say $10,000. This implies that a deck, a remodeling project, or other job that is modest in scope and cost requires less attention than a more complicated, more costly project.

But there is no reason to expect a poorly designed, undersized, overloaded frame to stand up just because it is simple and inexpensive. The joists on a $1,000 deck must be designed and installed just as carefully as the joists on a $100,000 house. Consequently, you should observe the following guidelines.

- Always seek professional help if you have any doubts about the structural design of your project.
- Always build within the limits of local codes. Avoiding them to save time and money, or to keep your home improvement off the tax rolls, is foolhardy. They provide an extra measure of protection over the designer's expertise (even if you are dealing with a registered architect) and are intended to eliminate the dangers of unsafe, untried materials and methods of construction.
- If in doubt, overbuild. This principle is inherent in most codes, and should carry through all stages of construction. If your span requirements, for instance, fall on the border between a 2 × 6 and 2 × 8, use the larger timber. Rounding off to the high side may cost a little more, it may even waste a little wood, but it may overcome a hidden defect in materials or an honest mistake in construction.

Wood is a very accommodating material. For the present, it is practical economically and ecologically. With the thorough application of proven structural principles, careful organization, and common sense, it will provide great strength for its size and weight, while remaining versatile enough to fulfill a wide range of engineering demands and architectural tastes.

FRAMING ALTERNATIVES

Up to this point, all the information on framing materials has been about wood, specifically dimensional lumber like 2 × 4's and 2 × 10's. But statistics on lumber in new, single-family houses show that wood timbers are not used throughout the house the way they used to be.

Figures kept for FHA-inspected homes

since 1959 show that lumber consumed for sheathing has declined in use by nearly two-thirds, as has lumber used for subflooring. It has been replaced by softwood plywood that costs less and takes less time to install. In fact, when the framing module is enlarged from 16 to 24 inches center to center, frame stiffness can be increased up to 70 percent by gluing plywood subflooring to the joists.

Dimensional timber is still the premier framing material used for over 55 percent of all floor frames, 80 percent of all exterior wall frames, and over 90 percent of all interior, nonloadbearing partitions. But in some areas, its use has declined more drastically. Since World War II, for example, concrete floor slabs have been used more frequently as the importance of a full cellar declined.

Cellar space was needed historically to store coal for domestic heating. But as the country switched from coal to fuel oil, and later to natural gas and electricity, coal-storage areas became a superfluous part of residential design.

The idea of a full cellar, though, had been ingrained in the perceptions of most home buyers as an integral part of a house. And the idea was so hard to part with, the full cellar outlasted its utilitarian design role, and became the rec room of the fifties and sixties. Do-it-yourself magazines and books of this era are filled with projects like boxing in the soil pipe, and partitioning off the laundry area, to make hard, gray, pipe- and wire-filled cellars more livable.

It took more than 20 years to get everybody back upstairs into what is now called the family room. But in many regions of the country, the full cellar persists, as a shop, laundry room, general storage area, playroom, or garage, even though it is a hole in the ground that, in many houses, needs a sump pump and cannot be finished with wood flooring or carpeting because it leaks like a sieve.

A joint FTC-HUD study of newly built houses, conducted in 1980, showed that 6 percent of the new homeowners had a serious complaint with the builder about water in the basement. This common problem is aggravated in older homes by extremes of seasonal variations, particularly in northern states, where deep frost causes heaving, and spring thaws saturate the ground with water.

Considering these factors, cellars should have disappeared from houses in northern states first. But the opposite is true. Concrete floor slabs and cinder block walls have replaced lumber framing in the South more than anywhere else. While the North hung on to the cellar past its useful design lifespan, and in spite of unfavorable environmental conditions, the South opted for as much masonry as possible becaue it provided better resistance to a warm, humid environment with its own special problems, notably lumber-eating insects and lumber-pulverizing rot.

Masonry is a good but generally more expensive alternative to wood but there is another that resembles structural wood framing more closely. It is a material that uniquely combines great compressive and tensile strength that has changed the shape, and, in turn, the character of cities, and that makes it possible to build structures a hundred stories high that can sway two feet in severe winds with complete safety—steel. Along with reinforced concrete, steel is the most influential commercial building material of the twentieth century.

Did you catch the key word, *commercial*? In a revised edition of this book, written in the early twenty-first century, the wood to steel proportions of this chapter might be reversed. Right now, though, steel remains a commercial product, with only a tiny share of the residential market.

Steel is vermin-proof, fire-proof, it won't rot, it won't warp, and, compared to wood, it does not deteriorate. With all these advantages, home builders should be using steel, and they might if they could get it. But steel is a large-volume material requiring heavy investments at the production end that can only be justified by large-volume marketing. Steel is not as accommodating to on-site variations as wood. It also requires considerable lead time from plan specification to delivery. These and other characteristics make steel very difficult to stock in lumberyards and building supply houses.

Two other factors keep steel from gaining more than a 3 percent share of the residential framing market. Steel construction requires tools and skills different (though not necessarily more complicated) than wood framing, and it is generally more expensive to buy and install, particularly in custom-designed, irregularly dimensioned houses, as seen on this table compiled by the U. S. Forest Service.

These figures show a sampling of the increase in building material prices that has put most new houses beyond the reach of most young home buyers. They also pinpoint one of the major drawbacks inherent in wood framing. Exterior, load-bearing walls need the strength of 2 × 4 studs. And this framing thickness must be consistent throughout the house to accommodate electrical outlet boxes, door frames, hardware, and other materials that are based on the 4-inch dimension of a wood stud.

	In-Place Costs of Wood and Steel Framing					
Year	*Floors*		*Partitions*		*Load-Bearing Walls*	
	Wood	Steel	Wood	Steel	Wood	Steel
	Douglas fir 2 × 10	2 × 8 18 gauge	Douglas fir stud	26 gauge stud	Douglas fir stud	20 gauge stud
	Dollars per square foot		*Dollars per linear foot*			
1970	0.51	0.60	1.41	0.98	1.64	1.86
1978	1.13	1.24	3.02	1.89	3.56	3.60

FLOOR AREA = 1,047 SQUARE FEET

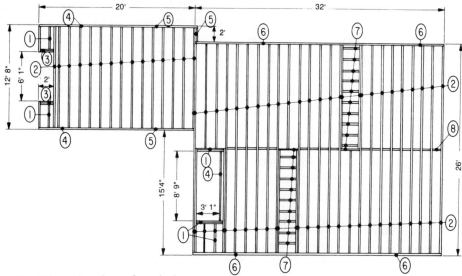

Figure 93. Schematic plan of typical wood floor frame. The circled numbers refer to different structural elements.

However, on partition frames (those inside the house that divide up the space without bearing heavy structural loads), the full strength of a 2 × 4 is unnecessary. Even though there is an obvious distinction between load-bearing and non-loadbearing walls, the same wood 2 × 4 stud is used in each instance.

The table shows how uneconomical this is, particularly compared to steel, where this distinction can be accounted for very economically. Where load-bearing strength is not required, the gauge, or thickness, of the steel in the stud can be reduced without altering the thickness of the wall. This gauge reduction brings strength in line with design requirements and substantially reduces the cost. This may be practical only in developments where a standardized, large-volume order can be placed well in advance of groundbreaking.

Practical Limitations of Steel Framing

Steel frame houses do not have to look like warehouses. They can be finished on the outside with conventional wood siding, and on the inside with Sheetrock. But these materials cannot be nailed on because there is no wood behind them, only hard steel.

Two methods are used to attach steel framing members to each other and finishing materials to the frame. One is a

FLOOR AREA = 1,047 SQUARE FEET

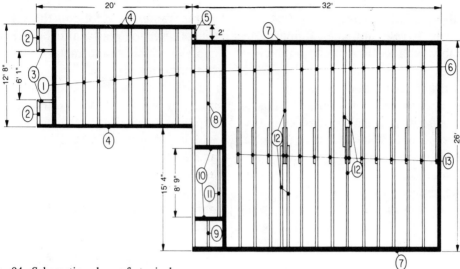

Figure 94. Schematic plan of typical steel floor frame. Less structural material is required when using steel than when using wood.

crimping tool, akin to a long-handled pliers, that pinches a steel joist and shoe, for instance, into a joint without using nails or screws. It's like dog-earing papers together instead of stapling them.

In place of nails used with wood frames, steel parts are joined with self-tapping drive screws, either directly or with an angled bracket that secures joints where pieces of steel meet at right angles. When materials are applied to the frame (Sheetrock, for example), an electric screwdriver is a necessity. The drive screws are sharply pointed to pierce the Sheetrock and the steel, while a torque-limit setting on the screwdriver stops turning them as the wide screw threads pull the panel close to the stud. This system does make very regular and very neat depressions in the wall surface that are easy to spackle and tape.

However, every time you want to hang a picture on the wall you will have to find a substitute for conventional, nail-carried hangers. And if you want to remodel (installing a window, for instance) you will need a power hacksaw. The imprints of steel studs and the fasteners attached to them also tend to bleed through an interior paint film more quickly and more noticeably than the imprints of wood studs with conventional nails. This is because steel has a negligible resistance to heat transmission. Not only will you lose more interior heat where the steel touches interior and exterior wall surfaces, but you will also have to repaint more often. The minuscule R-value for steel compares unfavorably with an R of 1.25 per inch of softwood timber.

During construction, steel may also cause problems with insulation (clips or tape are needed instead of staples) and require some extra work as piping and electrical cables are passed through joists, studs, and rafters. Steel framing members can be prepunched (although the holes may not always be where you want them), but since the material is a good conductor, plastic insulators should be inserted everywhere Bx cable (wiring wrapped in a spiral metal sheath) meets metal frame. This follows the safe practice of inserting plastic antishorts between cable conductors and metal sheath at all openings in a Bx line.

Steel may be the residential framing material of the future. It offers many advantages, including extreme durability, which will become increasingly important as maintenance and replacement costs on all housing components continue to rise. For now, though, you should become familiar with the practical limitations of steel before you or your builder place an order.

10

FLOOR AND WALL FRAMING

Much of the engineering information about wood framing members in the previous chapter represents a relatively recent change in residential construction. Houses used to be built solely by experience, with practices that gradually became standards as they were proven over time. With engineered construction, however, the properties of specific woods, grades, and sizes can be economically matched to particular frame designs.

If, for instance, commercial builders can find a way to achieve reasonable stiffness with $\frac{1}{2}$- instead of $\frac{5}{8}$-inch subflooring, for every five sheets needed they can use the equivalent of four instead. On a forty-house project with 1,200-square-foot homes requiring roughly 1,600 sheets of subflooring, they can do without the equivalent of 320 sheets of plywood. This kind of economy encourages contractors to build by the tables and not by experience.

Commercially efficient construction in many cases is a small step away from minimal construction. This kind of protection against the weather, site conditions, and other factors of deterioration may prove sufficient in average situations. It may not stand up under more severe or unexpected problems.

Many framing systems are suitable for residential applications, each with its distinctive advantages and disadvantages. Whichever one you use should certainly reflect efficient use of increasingly expensive building materials but not at the cost of durability. Furthermore, efficiency is not necessarily gained by reducing the amount of wood in the structure. Timbers have a collective efficiency determined by the way they fit together in the framing system. And in most houses this is the area that determines true efficiency.

The current phrase for this integrated planning is modular coordination or modular dimensioning. It can be applied throughout the frame as a single unit, independent of the flooring, wall, and roof frame; or, most frequently, to the floor and wall together and the roof independently. Floors and walls are flat planes. And although one is horizontal and the other is vertical, framing techniques are similar. You may not use the same size timbers in each frame, but it does make sense to use the same framing system.

The prime or major module is 4 feet, represented in plank and beam framing where heavy timbers are used 4 feet on center (notated o.c.), which means that it is exactly 4 feet from the center of one timber in a frame layout to the center of the next. The minor module is 2 feet, represented in so-called engineered systems using 2 × 6-inch wall studs to accommodate extra insulation. The most common module is 16 inches, dividing the 4-foot major module into three parts.

On floor framing layouts, substantial savings are possible by adhering to the 4-foot depth module. There is some waste with all dimensional timbers. A 16-foot 2 × 10, for instance (nominal size), which actually measures $1\frac{1}{2} \times 9\frac{1}{2}$ inches, is likely to be 16 feet plus $\frac{1}{2}$ to $\frac{3}{4}$ inch long. This trimming allowance is marginal. If the first-floor depth is 32 feet, for instance, two 16-foot joists, placed $1\frac{1}{2}$ inches inside the building perimeter to allow for a continuous, horizontal belt, may be lapped 3 inches ($1\frac{1}{2}$ inches from each side) over a central girder. Since dimensional timber is manufactured in 2-foot increments starting at 10 feet, a house depth up to 2 feet less than 32 feet would still require the 16-foot joists, creating waste. Similarly, a depth up to 2 feet greater than 32 feet would require the next greater size joist (in this case either two 18-foot timbers, doubling the waste, or one 18-foot and one 16-foot timber, which would halve the waste only if the girder was moved off center). Consider the examples at the foot of the next page for houses framed with a central girder.

As far as floor joists are concerned, modular house length is not so important. One more row of joists, or even an odd length out of the foot module will affect only the length of belt material, and will do it only once, not in multiples of the number of joist rows. However, the odd length size will be felt as dimensional siding, interior wallboard, and plywood subflooring are applied.

FRAMING SYSTEMS

The house superstructure, whether dimensional wood timbers, steel, truss-framed, plywood web-framed, or some other configuration, must serve many purposes. It must hold itself together as a unit, support the dead loads of building materials, the live loads of people who occasionally slam doors and jump up and down, and it must be able to move.

Framing systems are not provided with control joints as in masonry, because the materials they are made of, and the great number of joints connecting the independent elements, give the frame a limited but built-in flexibility. If the structure is too loose, wallboard nails might pop in a stiff breeze. If the elements are too rigid (this is almost impossible to accomplish, even if you try), connections between ma-

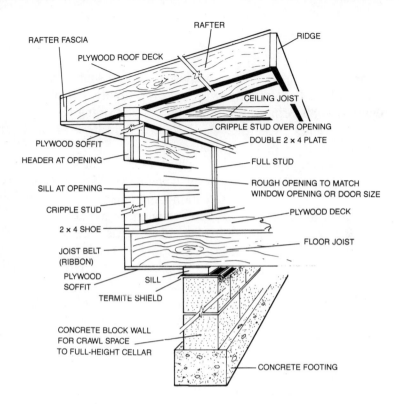

Figure 95. Typical residential frame components

terials might fracture under minor pressures.

Given this relationship, most framing systems are designed to achieve high strength. But at this point a distinction must be made between those materials applied to the frame that increase its strength, and those that depend on its

Original Floor Plan

22 × 48 feet—1,056 square foot area
2 × 8-inch joists, 16 inches o.c.
Required joist length—11-foot
Standard available joist—12-foot
Total joist rows required—37
Buy 36 × 2, 12-foot joists—888-feet
Total board feet—1,184

30 × 60 feet—1,800 square foot area
2 × 12-inch joists, 24 inches o.c.
Required joist length—15-foot (from 16-foot)
Total joist rows required—31
Buy 31 × 2, 16-foot joists—992 feet
Total board feet—1,984

Modular Alternative

24 × 44 feet—1,056 square foot area
2 × 8-inch joists, 16 inches o.c.
Required joist length—12-foot
Standard available joist—12-foot
Total joist rows required—34
Buy 34 × 2, 12-foot joists—816 feet
Total board feet—1,088 (8.1 percent less)

28 × 64 feet—1,792 square foot area
2 × 10-inch joists, 24 inches o.c.
Required joist length—14-foot (no waste)
Total joist rows required—33
Buy 33 × 2, 14-foot joists—924 feet
Total board feet—1,540 (22 percent less)

strength for permanence. For example, wallpaper in no way adds to the strength of a wall. But plywood subflooring does solidify a system of floor joists. Note the difference on the design table (page 151), between fiber stress in bending values for single joists and those tied together with subflooring.

It is good practice then to make a further distinction between pure structural materials (joists, beams, studs, and rafters), and applied structural materials like plywood subflooring over joists and siding over studs. Even though materials like these provide some measure of structural assistance, they provide yet another series of joints between dissimilar materials or at least different varieties of woods that absorb moisture, swell, and shrink at their own rates.

The best approach is to consider everything but the pure structural frame as a nonsupporting, decorative material that must rely on the frame strength for permanence. For instance, if floor joists bow in and out between perimeter belt and girder, don't rely on the plywood subflooring to stiffen them. Add cross bridging (or full-depth bridging made out of joist material), even if local codes do not require it. If stud wall corner posts can be racked even slightly out of plumb, do not rely on beveled siding to keep them rigid and square. Add let-in, 1 × 4, diagonal corner braces.

Think twice about any framing system that is made whole by a secondary application. And consider how each system relates to window and door openings on the plan. Fixed glass may be ordered to odd sizes (double glazing in anything but stock sizes is exorbitant), but most preframed windows and prehung doors do follow a modular plan (Figure 96). If openings are placed at random, or for architectural effect regardless of framing efficiency, filler studs may be required at frame edges in addition to modular dimensioned studs.

The central idea is to maintain the potential for complete structural soundness without needlessly overbuilding the frame. This potential is constantly under attack. At every stage of construction, there is a chance it may be diminished. You can think of the house as sitting on a natural spring that spews out problems. Solid structural systems put and keep a cap on the spring. But it is easy to develop leaks.

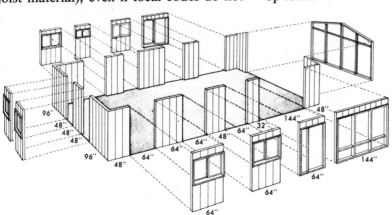

Figure 96. Efficient framing with modular windows and doors

Floor and Wall Framing

The spring may take the form of termites, attracted to wood debris buried in the foundation fill, that enter through a crack in the block wall. It may take the form of excessive energy costs due to air infiltration at corners that have racked, opening seams in the siding. It may take the form of constant interior maintenance as green studs dry and twist and warp, popping wallboard nails and opening taped seams.

Many of the most common complaints about houses can be traced to some inadequacy in the structural system, even though any particular system may fulfill one or more of its design functions thoroughly. For example, 2 × 6-inch wall studs may provide enough depth for energy efficient insulation batts but may drive up costs if filler studs are required.

Be sure to consider all the design implications of a framing system before plunging ahead with it simply because it appears, at least in terms of initial cost, to be the most economical.

16- and 24-inch Modular Systems

Both of these structural plans can be used with a 4-foot house depth module. Sixteen-inch framing was, until only a few years ago, the industry standard. Twenty-four-inch framing is the result of an increased awareness of energy conservation. In order to fit more insulation in the wall, the wall had to be thicker. As it turned out, the extra depth of a 2 × 6-inch stud provided enough additional strength so that one of every four studs in 16-inch framing with 2 × 4's could be eliminated.

Sixteen-inch framing is the smallest framing module, and because it interrupts the wall frequently and by small increments it offers the greatest number of choices for efficient window and door placement. With 16-inch o.c. framing, a 4 × 8-foot sheet of plywood siding (8 feet running vertically) would contact four structural timbers. Each vertical edge of the panel bears on half of a framing member. The two interior studs have full $1\frac{1}{2}$-inch bearing.

In order for each successive panel in the wall to have bearing, the 16-inch o.c. layout is carried forward from every stud except the first one in line at the corner. Since the panel must reach to the outside edge of this stud, not to its midline, $\frac{3}{4}$ inch (half the stud width) is subtracted from the initial 4-foot module. The convention is to hold your ruler at the outside edge of the corner, measure in 4 feet (the width of the panel), move back $\frac{3}{4}$ inch, make a square mark on the 2 × 4 bottom plate, called the shoe, and place an X ahead of the mark, working away from the corner. Studs will be placed on the X's, and squared against the lines.

Another simple method to obtain the same result (to allow for full coverage of the first stud while maintaining the 16-inch center-to-center layout along the wall), is to hold the ruler at the outside edge of the first stud, measure in $15\frac{1}{4}$ inches, make the square line, then move forward, called stepping ahead, to mark the X. Once the $\frac{3}{4}$-inch subtraction for center bearing on the fourth stud has been made, the layout may proceed 16 inches square line to square line, stepping

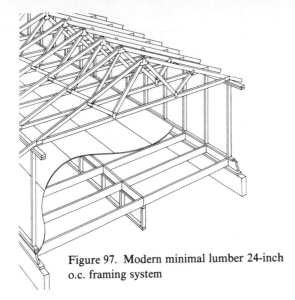

Figure 97. Modern minimal lumber 24-inch o.c. framing system

Figure 98. Material use is efficient in consistently modular frames.

FRAMING

ahead with the X mark each time. Remember to use a long tape for stud wall layouts to minimize errors, which are made likely by starting the ruler from scratch on each 4-foot module.

With 24-inch o.c. framing, each panel would contact only three studs, one along each edge of the panel, and one along its midline. A similar offset on the corner sheet is necessary here too. But because there are only three studs the first, offsetting dimension should be 23¼ inches.

Obviously, with only one central stud, siding panels are more likely to bow in and out on 24-inch framing than on 16-inch framing, where two inside studs are provided. This is less of a problem on rough floor and wall sheathing. For the most strength, 4 × 8-feet plywood panels are run with the 8-foot length at right angles to the framing members. Although there are more timbers to cross this way, the same offset rules are followed to assure bearing on the edge of the sheet.

Material thickness must be taken into account as well. Suppose ⅝-inch plywood siding is installed along one wall, starting at the outside edge of the first corner stud. This installation makes the stud wall joining it at right angles ⅝ inch longer. If this is not accounted for in the layout, two plywood edges will be exposed at each corner with a gap between them.

With both 16- and 24-inch o.c. layouts, it is profitable to view the layout marks you have made from the point of view of attaching materials to the frame. Take a sheet of plywood and lay it in place against test marks on the foundation sill. Make sure it will fall on the midline of a

Floor and Wall Framing

structural support before you nail joists or studs or rafters in place.

It is also important in each system to stick with the layout you start with. In other words, each rafter placed 24 inches o.c. will bear directly on a 2 × 6-inch stud placed 24 inches o.c., which, in turn, will transfer the load directly to a floor joist set at the same module.

The chief distinction between these systems develops from the differences in space between supports. At 16-inch o.c. intervals, ½-inch thick subflooring and sheathing is acceptable. At 24-inch o.c. intervals, however, stronger materials are required. Subflooring thickness is increased to at least ¾ inch, frequently in special grades (APA 48/24 plywood subfloor panels, for instance), and with tongue-and-groove edging to keep seams flush where panels meet at right angles to joists.

Generally 24-inch framing requires less labor but higher overall material cost. (There are fewer timbers to place and fewer joists, studs, and rafters to nail panels or siding into.) Floor area is the same regardless of layout module, but with 24-inch framing thicker and more expensive stiffening skins are required; while standard ½-inch plywood will bridge 16-inch spacing.

Data from the National Association of Home Builders Research Foundation have shown that material costs for wall framing are lower with 24-inch framing, while for siding (greater thickness needed between supports), they are higher. Combined material costs for wall frame and siding are roughly equal, though, in each system. Similarly 24-inch floor framing is less expensive than the 16-inch equivalent, but the subflooring costs are higher, again combining to produce a cost equivalence for both systems.

Overall, the test data produced material costs of $1,670 for 16-inch walls and floors, and $1,650 at a 24-inch module. When the cost of labor time was computed, however, total wall and floor costs were $2,650 at 16 inches o.c., and $2,490 at 24 inches o.c., representing roughly a 6 percent saving with the 24-inch module.

Western and Balloon Framing

Both these framing systems use dimensional wood timber in accordance with the National Design Specification limits for sizes, grades, and spans. Generally

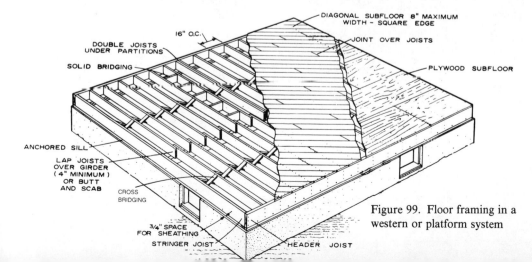

Figure 99. Floor framing in a western or platform system

they use the same amount of wood and require the same amount of labor. But they put the pieces together in different ways.

Western, or platform, framing is a staged process where each floor is framed and covered with decking before stud walls are erected. With this system studs are cut to length, set to a 16- or 24-inch o.c. layout on a 2 × 4 sole plate (the shoe running horizontally between the studs and the subfloor), and to a similar layout on a top plate (both horizontal members should be laid out at the same time to assure consistency).

This stud wall sandwich can be easily assembled on the finished subfloor platform. Simply stand on the 2 × 4 (resting on edge at right angles to the shoe), and drive two 16d galvanized common nails through the shoe into the end grain of the stud. If your weight is not enough to hold the stud in position against the hammer blows, tack temporary blocking on the platform for support. When the stud wall, complete with openings, is framed with shoe and single top plate, it can be tipped up into place at the edge of the plywood platform (see pages 198–201 for more about this procedure). Some contractors take the opportunity to install sheathing on the wall frame while it is on the deck. This may eliminate the need for scaffolding outside the building during framing. Tip-up construction is fast, and the walls produced are strong, due to the holding

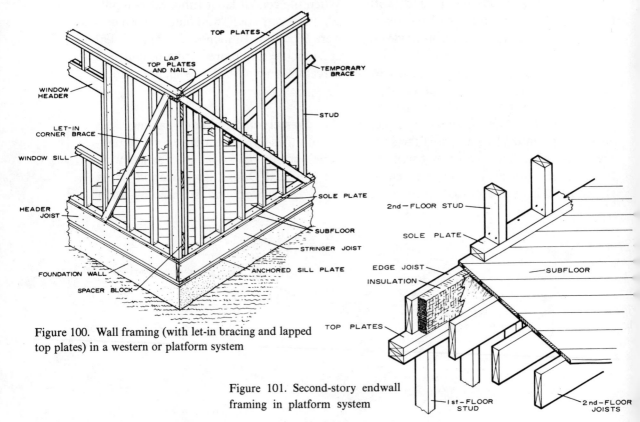

Figure 100. Wall framing (with let-in bracing and lapped top plates) in a western or platform system

Figure 101. Second-story endwall framing in platform system

Floor and Wall Framing

power and clean joints achieved with end-grain nailing. The alternative is called toenailing, or angling 8d common nails through the sides of a stud standing on the shoe. Holding power is greatly reduced, and the likelihood of splitting the stud by nailing too close to its edge increases.

With western framing, each stud wall is plumbed, braced, and locked in position by installing a second, horizontal top plate, lapped at the corners (Figure 100). Second-floor joists or rafters rest on the plate, and a new platform is built. This system has built-in fire stops (blocking that prevents extensive, in-wall drafts to develop), and equalizes vertical shrinkage within the frame (Figures 101 and 102).

The chief difference between western and balloon framing is in the outside walls. With balloon framing, joists are laid on the foundation sill, and instead of installing subflooring, wall studs, running full height from sill to rafter, are set next to the joist, directly on the sill (Figure 103). Both members are toenailed to the sill (three 8d nails each), and spiked to each other (three 10d nails, two from one side, one opposing). A final joist is face nailed to the studs running parallel to the floor frame to act as a fire stop, and to support the edges of subflooring. Individual blocking must be fitted between joists that rest on the sill adjacent to the full height studs.

Since no top plate is constructed at the

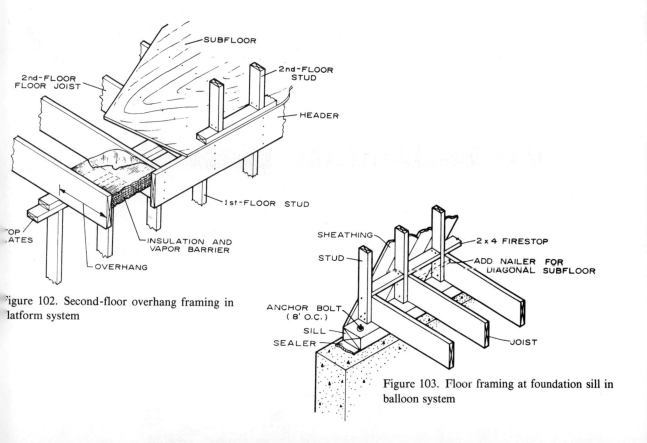

Figure 102. Second-floor overhang framing in platform system

Figure 103. Floor framing at foundation sill in balloon system

second-story floor level, joists at right angles to the stud wall are run into the full thickness of the stud, face nailed as on the first floor, and further supported with a ledger consisting of 1 × 4-inch boards, let in (mortised), to the inside face of the studs. Again, the last joist parallel to a stud wall is face nailed, while blocking is used between joists intersecting walls at right angles.

Diagonal corner bracing (1 × 4's let in on the exterior face of studs), running from the second-floor corner down to the first floor at a 45° angle, may always be used to add stiffness, but it is strongly recommended on balloon frames. This framing system provides minimal vertical movement, and, therefore, is frequently recommended for full, two-story construction, particularly with stone, brick, or stucco veneer.

Fire-stop blocking between floors also increases frame strength, as does corner bracing. It is good practice to install blocking on the closing side of all door openings as well. This ties two or more studs together to resist deterioration from opening and closing action, particularly on heavy, sliding glass doors. Consult local codes for bridging requirements, typically the same in both systems. Again, a stronger, reinforced floor frame will help keep finished flooring in place. Cross bridging (usually made of 5/4 × 3-inch material), has been widely replaced with solid bridging, made from material equal in depth to the floor joists (detailed on page 196). Typically, it is required at 8-foot intervals in unsupported floor joist spans. However, for kitchen, bath, and entryway floors that support ceramic or quarry tile, bridging may be installed at 4-foot intervals or, alternatively, joists in these areas may be doubled up (two joists spiked together), in order to minimize disruption of grout.

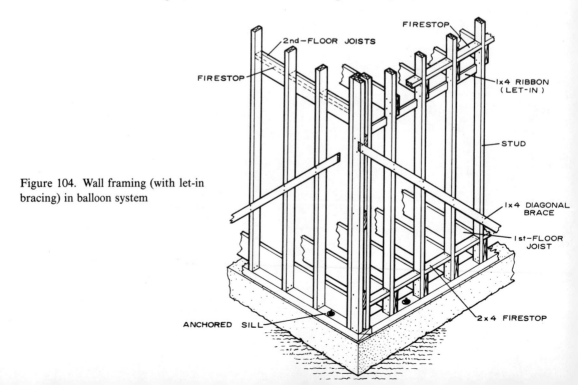

Figure 104. Wall framing (with let-in bracing) in balloon system

Floor and Wall Framing

Plank and Beam Framing

This is a very interesting alternative for residential framing, not widely used due to its distinctive design and both the size and weight of its elements, all of which makes it a difficult system to mass produce economically.

While western and balloon frames concentrate strength in the timber frame, which is augmented with subflooring, sheathing, and siding, plank and beam frames distribute the strength requirements more equally between the pure frame and applied structural materials. In this system, the pure frame consists of 4-inch wide (nominal), dimensional timber, tied together on a 4-foot center-to-center module with 2-inch-thick (nominal) structural planking. Each joist or rafter is roughly equivalent to two 2-inch-wide, conventional framing timbers spiked together. Each plank is roughly equivalent to a 2 × 6 or 2 × 8 framing timber.

Joists, depending on span and load requirements, might be 4 × 10's (massive timbers that need several workers for placement). The first floor would be built in one step, using 2-inch-thick tongue-and-groove planking, usually 6 inches wide. One side of the tongue-and-groove joint is beveled, which, when placed down against a 4 × 8 or 4 × 10 exposed rafter, presents a finished wood ceiling. The other side of the planking has square edges that form a butt joint for floor surfaces to be covered with strip flooring, carpeting, or tile with an underlayment.

Above each joist, one 4 × 4 (or two 2 × 4's spiked together), is set with 2 × 4 studs at 16-inch intervals between posts. A conventional double 2 × 4 top plate locks the wall, and each 4-inch-wide rafter rests on the plate directly over a 4 × 4 post.

Construction time is fast, although if the rafters are left exposed, and they usu-

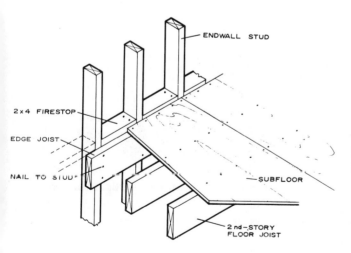

Figure 105. Second-story endwall framing in balloon system

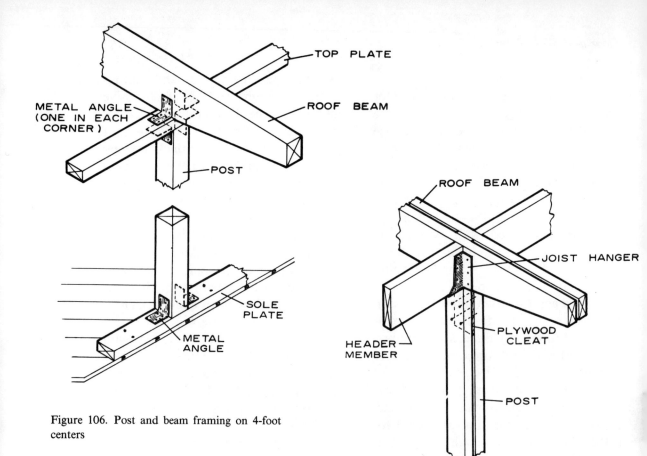

Figure 106. Post and beam framing on 4-foot centers

Figure 107. Post and beam framing with standard timbers sandwiched with spacer

ally are, care must be taken with their finish during placement. Once the timbers are in place, structural decking (available in long lengths by 2-foot increments), can be face nailed on rafters, or, on joists, angle nailed at the tongue with a floor nailing machine to hide nailheads. Joints are staggered on alternate rafters, or they can be end-matched (a tongue-and-groove joint is cut into the end grain of the planks), so that random lengths may be used and joined in rafter bays.

In addition to distinctive aesthetic characteristics (in a nutshell, rugged but modernly utilitarian), ceiling height is extended by the depth of exposed joists on the second floor. If the design is acceptable, if not pleasing, a profitable by-product of its simplicity is the combination of several construction stages and the elimination of some altogether. For instance, bridging is not normally required, although concentrated loads from water-filled bathtubs and food-loaded refrigerators may require additional framing if they will rest between joists.

Floor and Wall Framing

But every system has its drawbacks. With exposed plank and beam framing, wiring and piping that is hidden in conventional floors and ceilings must be boxed in. Insulation must be added in rigid panels (preferably foam for a high R-value per inch) above the roof planking, and supported by wire mesh or ledgers under the floor planking. Also, the lack of dead air between the conventional second-floor sandwich of finished ceiling and flooring materials encourages the transmission of temperature and sound floor to floor. Finally, the sheer size and weight of long 4-inch-wide timbers necessitates special handling and fastening (typically lag bolts and washers—see page 188), although beam-hanging hardware is available in 4-inch widths.

Specialty Framing Systems

Many proprietary framing systems are available now. As construction costs rise, there seems to be a place for new companies that have come up with a new way for builders to save money. Many of these systems incorporate a patented manufacturing process (a new way to splice or laminate beams, for instance), and a distinctive way to put framing members together.

Dimensional framing timbers are a

Figure 108. New proprietary framing systems with laminated lumber

Figure 109. Glued and laminated beam made of plywood sections

prime target for cost cutting, and several companies manufacture replacements in the form of truss timbers. Floor joists may be made of a top and bottom flange (1½-inch stud size or wider), connected by a web of single thickness, ⅜-inch plywood. Beams and headers for framed openings may be made of two 1¾-inch-thick pieces of plywood laminated together. Still other truss systems use conventional 2 × 4's laid on the flat, connected by strips of steel webbing.

Special frame systems like these are not for the small do-it-yourself addition. Availability may be limited. Code compliance may be questionable. If the local building inspector hasn't seen the system before, you may be in for extra inspections and delays. These systems are, after all, intended for volume use by professional builders who, in these times, are trying to cut back on material and labor costs wherever possible.

Other, less radical systems that do not involve a proprietary product or method are less risky. Plen-Wood (Figures 110 and 111), a relatively conventional framing method that uses a full, insulated

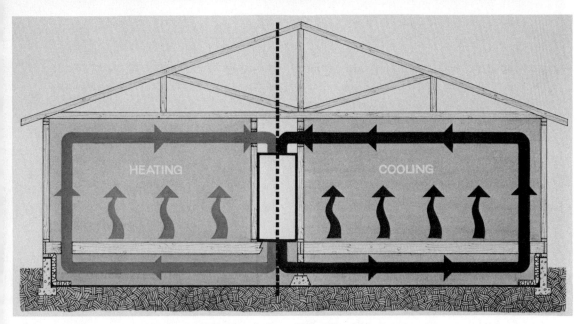

Figure 110. Sectional view of Plen-Wood framing system

Figure 111. Plen-Wood frame allowing distribution of conditioned air

crawl space as a giant heating and cooling plenum (no sheet metal ducts), has been explored by the NAHB and many lumber and plywood trade associations. And extensive literature is available on the system.

Remember, if you do stray from the ordinary, see how the local building department will react before you order any material. If you decide to use a proprietary joist, steel framing, laminated beams, or dome construction, for instance, get detailed construction and performance data from the manufacturer or appropriate trade association, coordinate it with your building plans, and take it all down to the inspector. Don't find out about code compliance after the fact. It can be expensive.

Steel Framing

Like specialty framing systems, steel joists and studs are almost entirely in the province of commercial builders, although not for use only in commercial and public buildings.

The local lumberyard will not have racks full of steel joists and studs. They may not even be able to order them for you. You will need to find a commercial building supply house or go directly to a steel company that has a construction subsidiary. The trick is to get them to sell you less than a freightcar load.

Most steel framing is three-sided. A joist, for instance, will have one vertical wall, ribbed at intervals for stiffness, with a top and bottom bearing flange. Connections between framing members are commonly made with self-tapping screws, set with a tool very similar to an electric drill, or, in the case of studs attached to a horizontal shoe, with a crimping tool that, in effect, dog-ears the pieces in position.

Several peculiarities of steel framing should be noted. For instance, access for piping and wiring is provided by perforations or punch outs in the steel. But at each hole, antishorts (plastic insulators) must be installed to prevent short circuits in metal, BX electrical cable from traveling into the steel frame.

Bridging is provided with continuous strips of thin steel woven over and under the joists. Another oddity, called ghosting (light gray shading that can develop on walls where they contact steel studs), can result from outdoor temperatures being transferred through the studs to the wall surface at a high rate.

Surprisingly, steel framing members are slightly lighter than their wood counterparts. But other advantages of steel over wood are not so surprising. Maximum allowable spans are roughly 10 percent longer. Steel frames will not warp, split, absorb moisture, swell, run sap, or burn. They are unaffected by dry rot, wet rot, and termites. All in all a high-strength, lightweight, extremely durable building material.

NAILS, LAG BOLTS, AND FRAME HARDWARE

Old-fashioned braced frames, the kind used in massive seventeenth- and eighteenth-century New England barns, did

Floor and Wall Framing

not contain a single nail. And many of these buildings are still standing, which does not say much for the importance of nailing. Massive, solid oak, hand-hewn timbers were cut with mortise and tenon joints (a peg protruding from one member fits into a hole cut in the other) and locked together with wooden dowels inserted at right angles to the joint.

On large timbers (6 × 6's and 8 × 8's), this durable but time-consuming method has given way to driving spikes—really giant common nails (7 to 12 inches long) that will rattle your bones if you mishit the head with a 2-pound hammer or sledge. They are generally used in predrilled holes for building railroad tie retaining walls.

Wire nails, called common nails, are designated in penny units, probably held over from a time when X number of pennies bought X numbers of nails. In any case, penny weight is notated with the letter "d." For example, an 8d nail is called an 8-penny nail. For each penny weight 2 through 12, length increases $\frac{1}{4}$ inch through 12d (3¼ inches). Above 12, 16d to 60d nails are calculated by $\frac{1}{2}$ inch per pennyweight.

The largest common nail is 6 inches long (60d), used primarily on commercial construction like pole barns where 2 × 4 purlins are nailed on edge to truss rafters. For 2-inch-thick dimensional framing, three sizes cover virtually every application: 8d common nails (2½ inches long), for toenailing and securing 1-inch-thick ledgers and nailers; 10d common nails (3 inches long), for face-nailing and spiking timbers together; and 16d common nails (3½ inches long), for spiking three 2-inch-wide timbers together, and, for extraordinary strength, face-nailing through one member into the end grain of studs or joists.

Good nailing requires the right nail type in the right amount at the right points for each connection. Failure of any

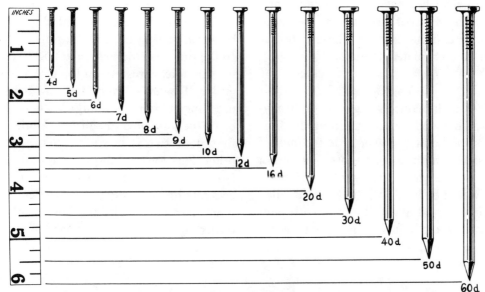

Figure 112. Common nail sizes and weights

one requirement will needlessly weaken structural connections. For instance, when diagonal 1 × 4 bracing is let into the faces of wall studs, the temptation is to place only one nail. But to get your money's worth from the let-in work and the lumber, architectural specifications call for two 8d nails into each stud, and four nails where the brace joins the corner post. Placing one, or in an attempt to gain extra stiffness, three nails is pointless as a single or center nail acts only as a pivot point under racking stresses.

In addition to size, number, and placement, nail finish should be considered as well. Wire nails are made of mild steel in cylindrical form that, compared to rectangular cut nails (for wood to masonry connections), have only moderate holding power. Nails not exposed to the weather may be bright (uncoated), while nails on decks, sill to joist connections, and the like, should be galvanized to prevent corrosion. If available, use hot-dipped galvanized nails (they have a rough and sometimes thorny coating), which are more durable and have more holding power than thinner, plated coatings.

Lag Bolts

A ¼-inch-diameter lag bolt, a heavy-duty, threaded bolt, has roughly four times the holding power of a 16d nail of the same length. Lags are used instead of nails where great strength is required, and would be hard to achieve with nails.

They should always be used with pilot holes to avoid splitting. For Douglas fir, southern pine, and similar woods, these pilot holes should be 60 percent of the lag diameter for short lengths, and 75 percent for 6 inches and up. The lead hole should make way for most of the shank diameter, but not for the threads, which bite into the wood. Ideally, the pilot hole is small enough so it is difficult to make the final turns to seat the lag on a washer, but not so small that you shear the lag head before it seats.

As a rule, the threaded portion of the lag should penetrate to a depth roughly ten to twelve times the shank diameter, and, when placed parallel to the grain (through a 2 × 10 belt into the end grain of a 4 × 10 joist, for instance), the center of the lag should be at least 1½ times the diameter away from the edge of the wood. In denser woods, in cold weather, or generally to reduce the size of the pilot hole, run the lag threads across a candle or bar of soap. This lubricates the turning and may prevent shearing the head.

Frame Hardware

There are large structural differences between bearing on, and butting against. In the first case the load of a 2 × 10 joist, for example, is resting on top of a 4 × 10 girder. However, every time a girder is used for support, framing depth is doubled (the joist plus the girder). Under the first floor this is not a problem, but in other locations it may use up finished space, create a low pass-through, or necessitate higher ceilings. Consequently, in many framing details the joist would be butted against a perimeter ribbon (structural banding using floor joist material

Floor and Wall Framing

Wire (common) Nail Data

Penny Weight	Length (inches)	Gauge #	Wire Diameter (inches)	Number per pound	Lateral Strength (pounds for Douglas fir)	Withdrawal Resistance
6d	2	11½	.113	167	63	29
8d	2½	10¼	.131	101	78	34
10d	3	9	.148	66	94	38
16d	3½	8	.162	47.4	107	60

Recommended Nailing Schedule

Joint	Method	Amount	Nail Size
Header-Joist	end-nail	3	16d
Joist-Sill and Joist-Girder	toenail	2	8d (10d optional)
X-Bridging-Joist	toenail	2	8d
Solid Bridging-Joist	end-nail	3	16d
Shoe-Joist (through subfloor)	face-nail	2, 16-inch o.c.	16d
Shoe-Stud (horizontal assembly)	end-nail	2	16d
Shoe-Stud (vertical assembly)	toenail	3	8d
Plate-Stud	end-nail	2	16d
Second Plate-Stud	face-nail	2, 16-inch o.c.	16d
Rafter-Plate	toenail	2	8d (10d optional)
Rafter-Ceiling Joist	face-nail	5	10d
Stud-Stud	face-nail	stagger	10d
Brace-Stud (let-in)	face-nail	2	8d
Plank-Beam	toenail	1 (at tongue)	10d
	face-nail	1	10d

NOTE: toenailing (nails driven at an angle through the side of a timber, through its base, and into the support) should be done with 8d nails on 2 × 4's and smaller members. Use of 10d risks splitting but can be accomplished on larger timbers. See location of nails on framing detail drawings.

and running perpendicular to the joist direction). Also, roof rafters could be notched to rest on a center ridge, or placed flush against it to incorporate ridge depth in the roof.

When joists and rafters do not rest on supports, they are held with nails or lags. But it is good practice to build in some form of bearing. This can be done with ledger strips, 2 × 4's nailed flush with the bottom of a 4 × 10 girder, for instance. When the 2 × 10 joist is placed, a 2 × 4-inch notch is cut so that 6 inches (nominal) of the timber butts against the girder and rests on the edge of the 2 × 4.

The same support can be provided with beam hangers, also called strap hangers. Many proprietary hangers are available in configurations to connect all sizes of 2- and 4-inch-thick timbers to girders, belts, rafters, intersections, and other joint combinations. Most have a pocket into which the joist or rafter is fitted, supported by nailing flanges that lie against the jointed timber.

For 2-inch timbers, 8d nails or 8d box nails (which have thinner shank to fit through prepunched holes in the hanger) are sufficient. This is not finish hardware and is typically used only when the mechanically supported joints will not be visible. Concealed hangers (brackets less than the timber width with central webs or pins extending into the beam or rafter), solve this problem but are difficult to find and time consuming to use.

In many situations you can make your own hanger, or at least gain strength by nailing on metal tie straps—1 inch wide,
thin steel pieces placed to bridge a wood joint. This is particularly useful on 4 inch rafters that are designed to meet a central ridge along the same 4-foot module. The first rafter placed can be lag bolted through the face of the ridge. However, this option is no longer available for the second rafter. In this case, steel hangers and tie straps, running at least 6 to 8 inches along the top of the rafter, across the ridge, and down the top of the abutting rafter, can be nailed on to reinforce the joint.

Remember that even when hangers are used for support, the connection must

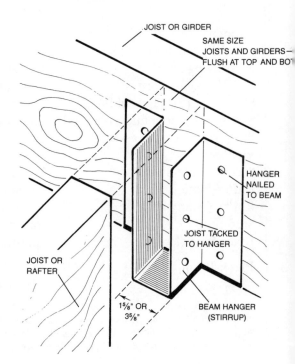

Figure 113. Typical beam hanger detail

Floor and Wall Framing

still be secured with good nailing or lagging. You may use hangers instead of a ledger strip to support joints between floor joists and perimeter belt, for instance, but each joist should still be end nailed.

FRAME CONSTRUCTION DETAILS

Posts and Girders

Vertical supports for girders—used wherever extra support is needed, as dictated by the span tables (pages 156–64)—must have pier footings independent of floor slabs. Since they carry heavy, concentrated loads, they need more footing support than 4-inch reinforced slabs can provide. Wooden 6 × 6-inch posts may be used to support first floor girders, typically three 2 × 8's to 2 × 12's, depending on loads, spiked together with staggered joints.

If wood posts are used they must rest on a concrete pier or steel pedestal (2 to 3 inches off the slab level), mounted on a steel pin set into the concrete (Figure 114; see also pages 57–58 on footing construction). The top joint between post and girder should not be made with 8d or even 10d toenails. Instead, make a connection that can withstand racking and twisting with two angle irons, one at each inside corner, secured with lag bolts.

However, to avoid shrinkage, compression, and the possibility of rot or termite damage, use cylindrical, steel lally columns, either filled with concrete and cut to length, or a hollow variety with an adjustable top fitting. Again, for proper bearing, the columns must be pinned at the floor, and secured to the beam with a flat mounting plate, which is usually built into the top fitting (Figure 115).

It is unusual for steel girders (typically I-shaped beams) to be used in residences. If they are, post dimensions must be shortened to accommodate a 2 × 6-inch

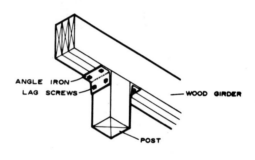

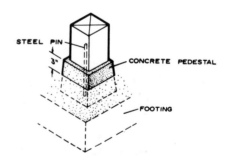

Figure 114. Wood post-and-beam girder with brackets and steel mounting pin

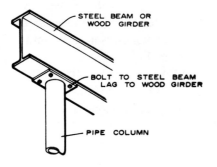

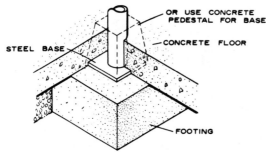

Figure 115. Steel post-and-beam girder requiring independent footing

sill, bolted to the I-beam top flange. Without it, you will have a tough time toenailing floor joists to the steel girder.

Make up wood girders on a secure surface, using dimensionally stable timbers. First, crown each board, i.e., sight down both 2-inch edges to determine which one bows slightly up in the middle of the span. Most dimensional timbers have a natural crown, although the amount of bow varies. Try to use those with roughly equal bows, all of which should be placed up. Spike the timbers together, first one to another with 10d nails, then the third using 16d nails from the opposite direction. Work from one end to the other, aligning the crowned edges as you go by manipulating the free ends of the timbers,

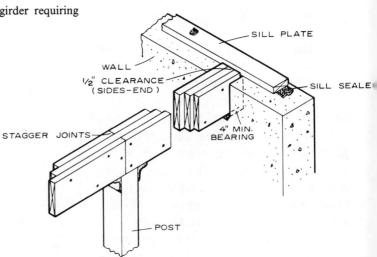

Figure 116. Built-up wood girder with bearing cut into foundation wall

Floor and Wall Framing

and, for fine adjustments, with 8d toenails through the 2-inch edges. Always align the top edges of the built-up girder; discrepancies from board to board are irrelevant along the bottom of a girder. Stagger joints as shown in Figure 116.

The finished, three-piece girder will be 4½ inches wide, and should have a minimum of 4 inches of bearing on the foundation at each end. Since the girder is frequently below sill height (if joists rest on the sill and the girder, the top of the girder must be flush with the top of the foundation wall), this bearing is created with a pocket formed in a poured concrete wall (see page 86) or a pilaster (built-out extension) on a block wall. To prevent rot or termite infestation (remember, the girder will be placed below the sill termite shield), pocket forms should be metal- or asphalt-lined (Figure 117).

Finally, for extra security, add 20d (4-inch) nails in pairs, from the side originally fastened with 10d nails, at 32-inch intervals. This built-up timber, centrally located, carries substantial loads. In many designs, bearing walls (walls that carry weight as opposed to partition walls that divide up interior space), are purposely placed directly over central girders. If there is any doubt about durability because of dampness, consider the use of pressure-treated timber.

Sills

Sills, which lie on the flat over poured or void-filled block walls, provide the masonry to wood transition. Place 2 × 6's, pressure treated for extra protection if de-

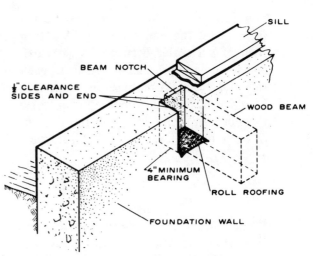

Figure 117. Asphalt- or metal-lined pocket to prevent beam deterioration

sired, with one edge flush with the outside edge of the foundation wall, over a sill sealer and a termite shield.

Sill sealers can be a continuous belt of fiberglass, asphalt, or other similar material that prevents air infiltration. A sill sealer is essential to energy-efficient construction because this seam is made between dissimilar materials that expand and contract at different rates.

Installation is straightforward. First, cut lengths to fit (apply preservative to the end-grain butt joints, if required), then lay the timbers in place to mark locations of anchor bolts (see page 82). Drill holes for the bolts ⅛ inch larger than the shank diameter, and drop the sill in place over the sealer and termite shield. Use the largest diameter washers possible for best bearing and to minimize cupping or warping. Lock them in place with nuts on the threaded rods.

At corners, 2 × 6's may be butted (again, seal the end grain, or fit a thin strip of sill sealer between the boards), and toenailed with an 8d nail from each side of the joint. A more stable corner can be made by lapping the 2 × 6's. To do this cut a ¾ × 5½-inch half mortise out of each board, creating a full thickness 1½ × 5½-inch lap.

Joists

Select floor joist species, grade, and size from the lumber design value and properties tables. Then lay out the sill for either a 16- or 24-inch o.c. spacing depending on frame design. With plank and beam systems, 4 feet o.c. must fall on one 4 × 4 or two 2 × 4's. It is important at this stage to plan for material thickness, and to remember the step back on the first 16- or 24-inch space to accommodate the absence of center bearing at corners (see page 175). Also, given a total wall thickness of stud, plus sheathing, plus siding, plus wallboard or paneling inside, it is unlikely that this module will work for interior and exterior materials. Siding that is applied from the outer edge of a corner post will be starting its 4-foot module several inches beyond interior wallboard that is started from the inside of the same post. Generally, hold the module for exterior materials, as wallboard, which comes in very long sheets, may be laid horizontally across wall studs.

For balloon frames, joists cover the full sill depth. On the more common platform frame they are placed to allow 1½ inches for a continuous perimeter belt (also called a ribbon or header joist). Tack each joist in place with an 8d toenail from each side (all crowns up), aligning the stringer joist (the last joist on two sides of the frame, parallel to the run), with 8d or 10d toenails into the sill. Then, to assure that the joists are plumb and to resist twisting, copy the joist layout on the ribbon (header joist), using a square to make plumb lines across the entire nailing face. Secure it to the sill with toenails between joists, and with three 16d end nails to each joist.

Joists may be butted and scabbed (reinforced with ½-inch plywood gusset plates, full depth, covering at least 6 inches past the joint on each joist), but it is better construction to lap the joints at least 4 inches. Remember, though, this will move half the joists 1½ inches off the framing module, and you will have to compensate by cutting back plywood subflooring. This would not matter if 1 × 6 tongue-and-groove, diagonal subflooring was used, although its cost and labor time have made it an uneconomical choice compared to plywood.

A final design note: Because the girder and floor joist system is responsible for so much of the house load, and because many maintenance problems stem from excessive movement in this part of the frame, you should attempt to minimize movement caused by uneven shrinkage between joists and girders. This may be done by using the same total depth of wood at girders (girder plus remainder of joist resting on ledger), as at the sill (joist plus sill).

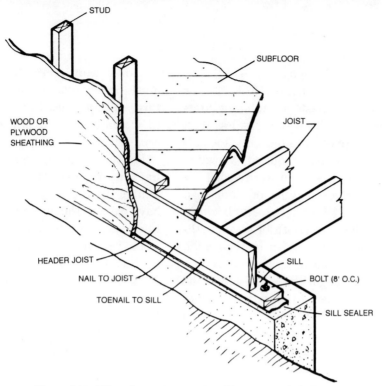

Figure 118. Sill sealer under wooden sill prevents air infiltration.

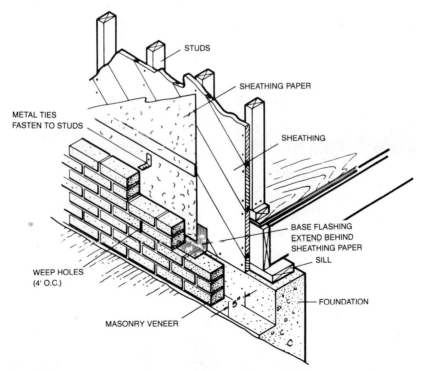

Figure 119. Joist length must accommodate wall materials and thickness.

Bridging

Cross bridging, generally 5/4 × 3 or 5/4 × 4 inches, is fastened at 8 foot intervals or at the midpoint of spans less than 16 feet. Because of the angle cuts, toenailing, and overall time required, however, cross bridging has been superseded by solid bridging, or lengths of floor joist material, cut 14½ inches to fit between joists 16 inches o.c. (Both of these methods of bridging are shown in Figure 99, page 177.) All solid bridging may be end nailed by snapping a chalk line across the floor, using a square to mark plumb lines on each joist, and placing the bridging pieces in a staggered pattern, first on one side of the line, then on the other. End-nail with three 16d nails.

Cantilevers

The overriding principle of cantilevering is to secure ⅔ to project ⅓. But this is a maximum. When exterior bearing walls rest on a cantilevered floor frame, loads must be computed carefully. Normally this projection, used for a bay window or to increase house depth, should not exceed 2 feet.

For a typical bay window extension in line with the joist run, simply extend and double up the joists at each end of the bay, extend the interior joists in single thickness, and tie them with a conventional belt or header joist, using three 16d or 20d nails per joist minimum (Figure 120). Since this construction interrupts the continuous perimeter belt, solid blocking must be toenailed between joist extensions, on the sill, in line with the belt on each side of the window extension.

For a similar window extension projecting at right angles to the joist run, the belt and the joists at least up to 4 feet inside the foundation (4-foot minimum pinned to support a 2-foot cantilever)

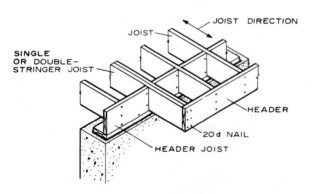

Figure 120. First floor overhang with extended joists

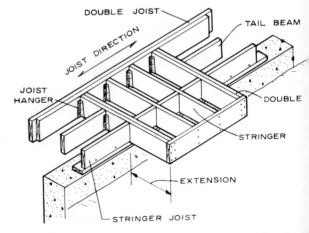

Figure 121. First floor overhang with frame pinned to doubled full joist

must be interrupted (Figure 121). The last full-run joist should be doubled, then the cantilever platform may be built as in the parallel plan. Other full-run joists should be nailed to the doubled cantilever joists on each side of the opening. This nailing is simplified by building the cantilever box in single thickness, end-nailing these interrupted joists, and finally spiking on the second joist that doubles the outside cantilever beams. Remember that this area will be exposed to the weather. Take precautions with extra insulation and a vapor barrier, closed in with exterior-grade plywood to form a soffit.

Floor Openings

Openings in the floor system for fireplaces, flues, and stairways (leave a minimum of 2 feet, 8 inches of clear tread) are treated like cantilevers. Joists that are interrupted by the opening are end-nailed to a header, which is doubled up. The floor load carried on these interrupted joists is then transferred to the headers, and in turn to the full-length joists framing the other two sides of the opening. To bear this extra load, these full-length joists should be doubled as well (Figures 122 and 123).

Remember to frame the opening 1½ inches larger than required at each header end. This permits end-nailing on the interrupted joists. Then the second header is spiked on after end-nailing.

Stud Walls

Stud wall framing is perhaps the most straightforward in the house. Balloon

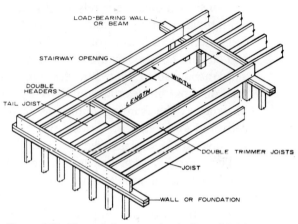

Figure 122. Framed floor opening in line with joist run

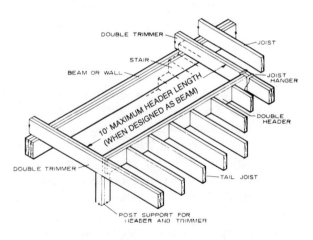

Figure 123. Framed floor opening parallel to joist run

frames (single, full-length studs), do require more elaborate bracing during construction, and for this and other time saving considerations, are rarely used today.

Platform framing (western, or tilt-up, construction), is used on development and custom homes because it is adaptable and manageable. Platform framing is used on those houses that you drive by on Monday as the framing lumber is being delivered and again on Friday when the windows are being trimmed.

The assembly sequence is fast and allows a choice of wall height, although it must be coordinated with siding and interior wallboard or paneling. Usually, ceiling height is 8 feet, although 8-foot studs plus a shoe plus two plates will yield 8-foot, 4½-inch ceilings. Shortened studs, called precuts, may be ordered as an alternative.

Construction speed is achieved by cutting all shoes and double plates to length first, remembering to account for material thickness. For example, consider a rectangular house with corners labeled 1 to 4, working clockwise. Shoes and first plates for walls 1–2 and 3–4 are full length. Shoes and first plates for walls 2–3 and 1–4 are cut 7 inches short of house width, allowing for two widths of stud wall shoes. Next, to lock up corners, allow for material thickness of the top plates on opposite walls. For example, cut the top plates for walls 1–2 and 3–4 less the 7-inch allowance, and top plates for walls 2–3 and 1–4 to full length.

Lay the shoe and first plate side to side, 3½ inch sides down, on the building perimeter. Test the measurements to make sure overlaps at corners are cut correctly, and align the shoe and first plate exactly. All this is done to align square marks that will guide each stud to a plumb position. It's a lot easier and faster than plumbing each stud with a level.

After this kind of careful preparation, walls will take shape faster than you imagine. Assembly is a simple, straightforward, repetitive process. Remember to lay out on center, subtracting half the stud width for full material coverage at corners. From this first reduced spacing, lay a square across shoe and plate at 16- or 24-inch intervals, placing an X for stud location on the right side of the line. This is certainly prejudiced against left-handers, but the convention is to work left to right.

Next, lay studs on the subfloor, on edge, roughly in line with and at right angles to the marks. Then move the top plate to the opposite ends of the studs, and end-nail through shoe and plate, making sure that the studs are aligned squarely with the marks.

Window and Door Openings

At this point, window and door openings should be framed, again while the wall section is on the deck, to permit end-nailing and eliminate less secure toenailing. The principles of construction are the same on walls as on floors. Loads are shared by studs in the wall. If you interrupt them with an opening, these loads must be transferred with a header to doubled studs at each side of the opening.

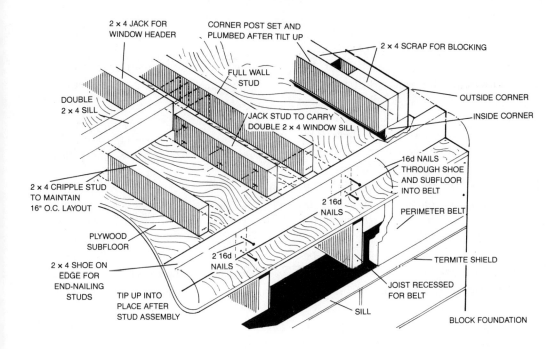

Figure 124. Stud wall frame, ready for end-nailing and tip up

Common sense dictates that if only one stud is interrupted the load transferring header could be a 2 × 4—load for load, size for size. But if three or four or more studs are interrupted the header must have the capacity to carry three or four times the load. It's logical but a surprising number of new homes have skimpy headers. It's the old song—reduce wood content and you reduce costs. But there are other by-products of this kind of false economy. Doors hinged to a single twisting stud behind the jamb will work out of alignment. Double-hung windows will bind in their tracks. Heavy sliding-glass doors will rattle the wall, pop nails, and fracture wallboard seams if they are closed vigorously.

There is one solid, long-lasting way to build stud wall openings. And this process is worth investigating in some detail as the procedures are common to framing throughout the house.

1. Determine the rough opening, supplied by the manufacturer of the window or door or vent, which is generally the complete framed unit plus $\frac{1}{2}$ inch shimming space.

2. Working away from the opening on each side of the sill mark a) the rough opening line; b) an O (the field symbol for a jack stud); c) a second square line $1\frac{1}{2}$ inches away from the first; and d) an X (the field symbol for a full stud).

3. Nail a full-height, shoe-to-plate stud over each X.

4. Establish the rough opening height with square lines across these full studs at the top of the opening.

5. At the bottom of the opening make

square lines 1½ inches lower than the rough opening. This space will be filled with a 2 × 4 sill (1½ inches deep). If a double sill is used—not a bad idea for large, heavy windows—make these lines 3 inches lower than the rough opening.

6. Cut cripple studs to fit between the shoe and the low mark on the full-height studs, enough to place one at every on center mark, and one against each full stud. Nail them in place.

7. End-nail a 2 × 4 sill, fitted between the full-height studs, with two 16d nails into each cripple stud.

8. Cut two jack studs equal to the height of the rough opening, place them on the sill against the full-height side studs, and face nail with 16d nails in a staggered pattern.

9. Make up a header (see depth table on this page), from two 2-inch-wide timbers sandwiching a ½-inch spacer, usually plywood. This assembly will be 1½ plus ½ plus 1½ inches thick, to equal full wall depth.

10. Place the header on top of the jack studs, and end-nail through the full-height studs. If more than 6 inches is left between the top of the header and the 2 × 4 double plate, maintain the on-center module by fitting in stud blocks.

That's it. Ten steps to make any opening where structural loads are not carried on nails, but transferred into framing members. Double sills are optional, but full-thickness headers, jack studs broken only by sills and headers, and, most important, headers of substantial depth, are not.

Header Depth Guide

Maximum Horizontal Opening (feet)	Minimum Header (with ½ inch middle spacer, inches)
3½	two 2 × 6's
5	two 2 × 8's
6½	two 2 × 10's
8	two 2 × 12's

This same procedure is followed for door openings, with the obvious omission of the sill and lower cripple studs. The full-height studs at each side of the opening rest on the shoe. But the shoe between these studs is removed, and jack studs run between the header and the subfloor.

Corners

Framed walls should be assembled, erected, and braced without corners. Their position is critical. They must form a plumb line between two planes, and nailing surfaces inside and out for modular materials. Because they represent a concentration of critical dimensions, they are fabricated on the deck, fitted into place, and nailed to the shoe and first top plate. It is good practice to verify every aspect of wall framing before locking up construction with the lapped top plates.

Two types of corners are suitable. The strongest but least economical is made by spiking together a sandwich of two full-length 2 × 4's, separated by 2 × 4

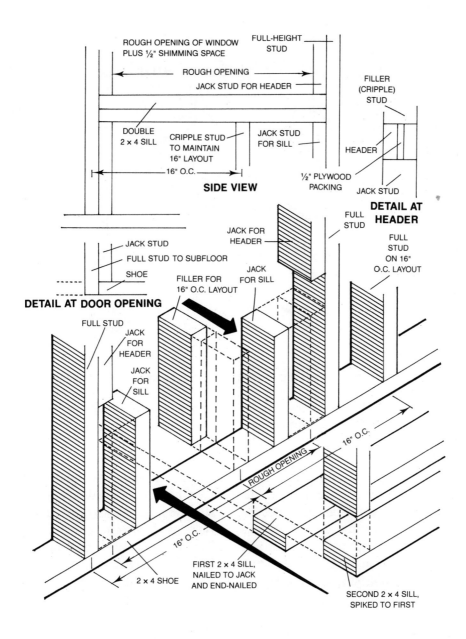

Figure 125. Typical framing at stud wall window opening—3/4 view

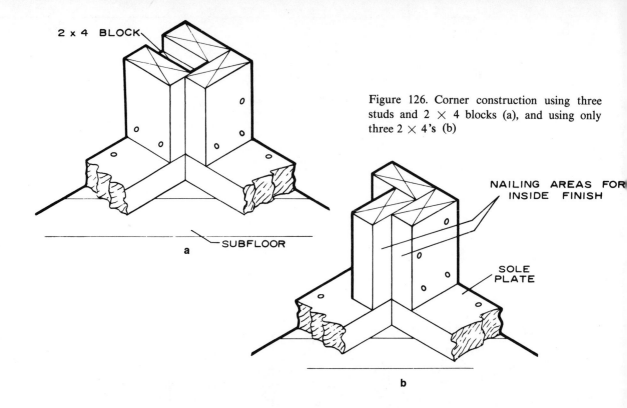

Figure 126. Corner construction using three studs and 2 × 4 blocks (a), and using only three 2 × 4's (b)

blocks—a nice way to use up scrap lumber—(Figure 126a). This assembly measures 4½ inches. (Three separate 1½-inch edges) by 3½ inches. Place the 4½-inch edge down on the deck, and spike a third 2 × 4 flush with either outside edge. The final 2 × 4 should cover one full stud, the blocked void (roughly 6-inch blocks placed at the top, bottom, and center will suffice), and ½ inch of the other stud. This creates nailing surfaces inside and out.

An alternate method of corner construction consists of nailing two full-length studs together into an L-shape, then adding a third stud flush with the base of the L, against its long side (Figure 126b). This method creates a good inside corner but an unbalanced outside corner. One surface provides 5 inches of bearing while the return surface provides only 1½ inches.

Keep in mind that solid framing is determined by the method and care of construction, but also by the quality of materials used. Moisture content (see page 146) must be 19 percent or less, preferably 15 percent or less. Select the framing timbers carefully. Pick through piles in the lumberyard if you must. Examine the grade stamps. Return heavily warped, bowed, knot-laden, and waterlogged timbers. You are paying for the goods. Sometimes, particularly if you have hired a contractor, you may underestimate the leverage this gives you.

11

ROOF FRAMING

There are two things to bear in mind about framing the roof of a house. First, you have to use a system compatible with the floor and wall framing. Rafters should fall over studs for proper transfer of loads. Second, you have to do a lot of the carrying and measuring and nailing work on rafters that are well up in the air. This literally adds another dimension to the job that might make you hesitant about climbing around on ladders, scaffolding, and framing members a story or more off the ground.

Everyone should have a healthy fear of heights. It encourages you to exercise caution as the consequences of a fall become more dear. But if this fear gets out of hand, it can impair your ability to work and actually make an accident more likely. It's simple—if the height really bothers you, don't go up there. Have someone who is more experienced and more at ease do this part of the work. And this goes for contractors as well as novice do-it-yourselfers.

ROOF DESIGNS

Many roof configurations can be used: plank and beam; flat; gable end; shed; hip; trussed; all at varying slopes, some with valleys and dormers. But in each case secure rafter placement depends on solid, square top plates.

The first step in all but plank and beam systems is to place the ceiling joists, following the steps for floor joists with only a few differences. Depending on roof slope, the top outside corners of ceiling joists may have to be trimmed flush with the angled rafters. They can be cut in advance, but it is good practice to trim them flush with the rafter by hand to double the amount of frame bearing along the roof edge.

Also, nailers must be provided between joints along wall-to-ceiling corners. These joints, whether plastered or Sheetrocked, are among the most difficult to preserve without maintenance. Framing stresses and strains rarely pull ceilings apart in

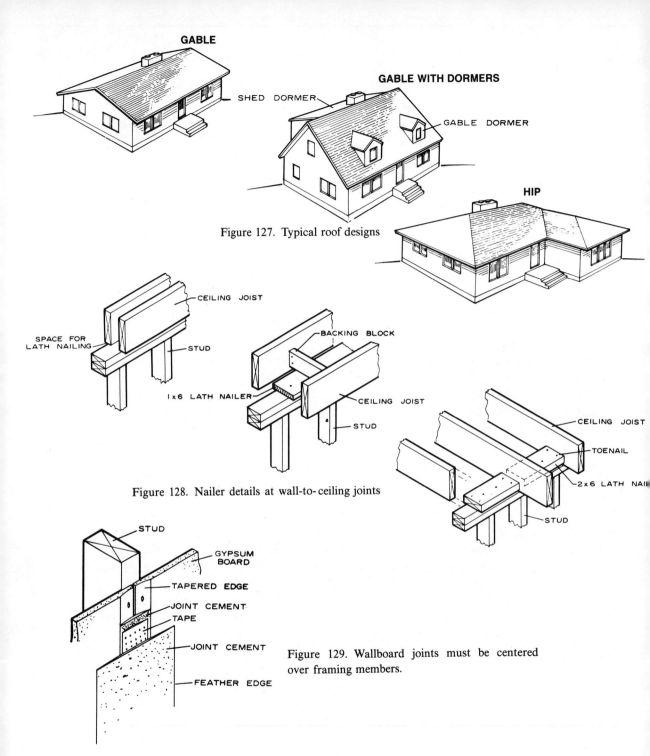

Figure 127. Typical roof designs

Figure 128. Nailer details at wall-to-ceiling joints

Figure 129. Wallboard joints must be centered over framing members.

the middle of the room. They tend to surface at the weakest link, the connection between two planes. Consquently, do not underbuild these nailers with the 1-inch or even 5/4-inch-material that is frequently recommended.

Ideally, 2 × 6 blocks should be placed flush with the outside of the stud wall, extending 2 inches into the ceiling area. This calls for a lot of material, but the ratio of $3\frac{1}{2}$ inches pinned to the plate, and 2 inches for nailing the edges of wallboard provides a very secure corner seam. However, 2 × 4 blocks with 2 inches pinned to the plate, and $1\frac{1}{2}$ inches extending into the ceiling area will suffice if well secured with gently staggered 16d nailing.

When roof trusses are used (see page 217), and this is now common practice in single-family and attached-housing developments, the bottom chord (the lowest horizontal, full-length member) serves as the ceiling joist. Trusses, available in varying slopes as well, are triangles. Without the bottom chord, the two legs of the triangle would tend to flatten out, pushing the sidewalls as they collapsed.

Withstanding this stress is another function of ceiling beams. They serve to tie the house walls together, a function performed by exposed collar beams on cathedral ceilings. To hold ceiling material and resist lateral loads, ceiling joists should be secured with three 8d or 10d toenails into the top plate and, for high strength, with beam-hanging hardware. Toenails should be driven wherever the joists cross partition walls as well (Figure 130). The more you can tie the frame together, the stronger it becomes.

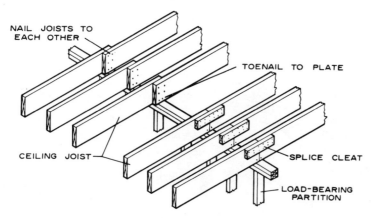

Figure 130. Overlapped and scabbed joints of ceiling joists over bearing walls

Plank and Beam Rafters

This 4-foot o.c. system is used with insulation above the roof decking to increase ceiling height, and expose 4-inch-wide dimensional beams. Rafters can rest directly on 4 × 4 posts (the joint should be reinforced with strap ties or beam hardware), or in line with the posts on a top plate. And if the top plate is a 4 × 6-inch header, the resulting space between posts (3 feet, 9 inches wide by stud height) becomes structurally independent of the frame. For example, this space could be framed for full, fixed-glass panels.

If rafters rest on top of the header, the ends of the bays between them must be closed with blocking, toenailed to the header and each rafter. If the header is run in line with the rafters, that is, at the

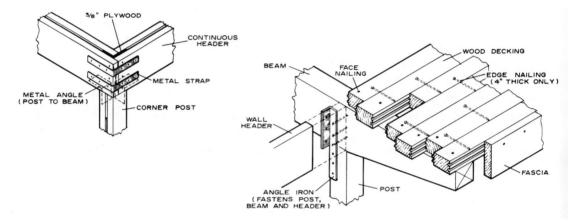

Figure 131. Hardware-reinforced connections for post-and-beam rafters

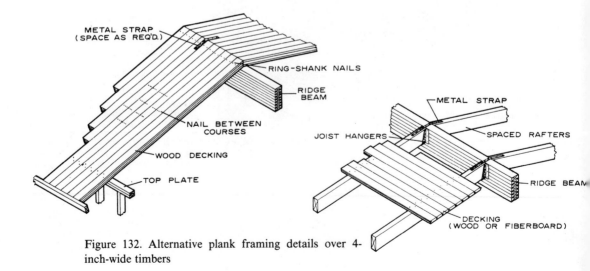

Figure 132. Alternative plank framing details over 4-inch-wide timbers

Roof Framing

same elevation on the wall, the 3-foot, 9-inch sections butt against the rafters and must be toenailed. In this case beam hardware, or the overbuilt detail of adding header-supporting jack studs on each side of the posts, is required.

Flat Roofs

Flat roofs are extremely easy to frame because the ceiling joists act as rafters. But they are difficult to weatherproof, and, in anything but a moderate climate, frequently a nightmare to maintain. Built-up roofing (several layers of asphalt roofing sandwiched with hot-mopped tar), is the most expensive, asphalt-based roofing system. It is difficult to drain without puddling. Drains clog easily. There are many design problems and not many advantages.

True flat roofs must be built up along the perimeter with a cant strip, a triangular strip about $1\frac{1}{2}$ to 2 inches high at each leg, laid with the hypotenuse toward the roof interior. Flashing is placed over the strip, which keeps water on the roof (in winter freeze-thaw cycles this can leave 2 inches of solid ice on the roof), until it reaches a drain outlet.

If flat lines are required for appearance, it is generally worth the trouble to extend the wall height (about 6 inches on most spans), to create a horizontal facade, while sloping the roof (at least 4 inches every 10 feet) behind it just enough to make standing water drain.

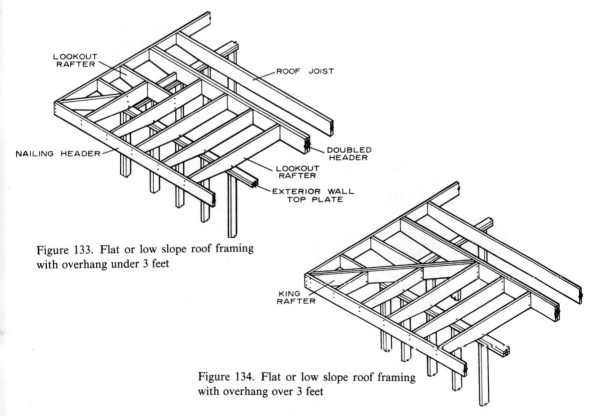

Figure 133. Flat or low slope roof framing with overhang under 3 feet

Figure 134. Flat or low slope roof framing with overhang over 3 feet

Shed Roofs

Also called lean-to construction, this is simply a sloping plane angled between two walls, one of which is higher than the other (see Figure 127, p. 204). Rafters that overhang the wall are notched with a cut called a bird's mouth to provide horizontal bearing on the top plate.

It is easily framed, flashed, and covered with roll roofing or strip shingles that carry surface water to a full-length gutter. When used in combination with another lean-to panel, modern architectural effects can be achieved. Shed roof design is complementary to plank and beam framing, which encompasses rafter, interior ceiling, and roof decking in the framing system.

Gable Roofs

A lean-to roof is essentially half of a gable roof, the design used most frequently today, and one that contains almost every roof frame detail that may be encountered, including ridge pole and top plate cuts, valleys, shed and gable-end dormers, and gable-end framing. This roof system uses rafters all cut to the same length and pattern. Once you make a template rafter, and determine that it fits correctly between plate and ridge, every other rafter can be cut from this pattern, presuming that walls are square.

Mansard design (steeply sloping roof panels connected with a low-slope, pyramid cap), and gambrel design (the classical two-panel barn design), are both variations that use gable framing details.

Hip Roofs

This design consists of four triangular panels with the wide ends over the outside walls, and the points meeting at the highest, center point of the roof (see Figure 127, page 204). Obviously, a full hip roof is mathematically possible only on a square floor plan. Therefore it is more common to find a modified plan where

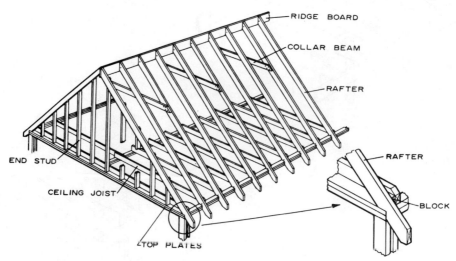

Figure 135. Typical gable roof framing

Roof Framing

hip panels from the short sidewalls of a rectangular house are separated by a ridge pole.

On hip roofs, few of the rafters are cut the same way. Every framing module (16 or 24 inches o.c.), requires a different length rafter between the outside wall and the angled hip rafter. Additionally, each jack rafter (those less than full length) requires a compound cut, that is, a combination of an angle on a vertical plane that accounts for the slope, with an angle on a horizontal plane that accounts for the out of square joint between wall and hip rafter (Figure 136).

This gives you two chances to miscalculate each rafter cut, dictated by finicky circular saw settings. And minor discrepancies force an unpleasant compromise: If the compound cut is too fat (leaves too much wood on the jack rafter), or too scant (too much wood is removed), the jack rafter will be pulled out of the framing module in order to make a secure joint. The temptation is to fudge the joint, not drive the nails home, or pound them in to seat the jack rafter against the hip rafter, or simply to throw the piece away and start from scratch.

ROOF CONSTRUCTION DETAILS

Once the style of roof frame has been decided, remembering that it must be structurally compatible with the floor/wall system, the next step is to determine the slope and therefore the length and size of required timbers.

Rafters slope is expressed by the inches of vertical drop per 12 inches of horizontal span. This is the convention specified on plans, for the limitations on shingles and other matters. If not stated specifically on plans, although it usually is, you can determine these numbers by measuring the distance between the top wall plate and the ridge (called the rise), and the distance between the midline of the roof and the inside of the perimeter wall (called the span).

The inches of vertical drop (the rise) per horizontal span gives the slope: for example, 4 in 12 (a $\frac{1}{3}$ slope), or a 5-foot, 4-inch rise, and a 16-foot span. Unfortunately, this tells you nothing about the length of the rafter. To find true rafter length you must first compute the load requirements of the span using the design tables. (See pages 156–64.)

Having determined the rafter size to satisfy load and horizontal span requirements (remember, the tables give span

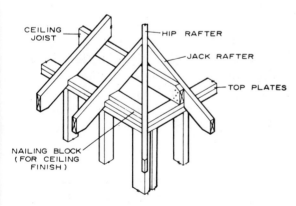

Figure 136. Rafter and ceiling joist construction on hip roofs

limits as horizontal, not sloping, distances), you can then use the Rafter Conversion Diagram, Figure 92 (page 155) to convert horizontal span to true rafter size.

This diagram shows the relationships of three factors: sloping distance (the true rafter size), horizontal distance (the length between roof ridge and sidewall), and rise (expressed in the inches of vertical drop per 12 inches of horizontal span). Any one value can be found if you know the other two; span and rise from blueprint measurements, for instance.

As an example, the horizontal distance from the center of the roof to the sidewall is given at 20 feet. The roof slopes down 8 inches for every 12 inches of horizontal distance. To find true rafter size, locate the intersection of 20 (horizontal distance) and 8 (slope in 12 inches), and follow the intersecting arc to find sloping distance, 24 feet.

Remember to add the length of wall overhang to this figure. In this case, a 1½-foot overhang would require a 25½-foot rafter—quite a handful. If length is either impractical or unmanageable, rafter length may be altered by decreasing the slope. In this example, if slope is 4 in 12, the 4/20 intersection shows a 21-foot true rafter size, or 22½ feet with the overhang.

Rafter Layout

In order to maintain the floor/wall framing module, rafters should be laid out the same way studs are measured for the wall, that is, accounting for full material coverage in the first module (see page 175), and marking an X ahead of each square line. In the same way that floor shoes and first top plates are marked at one time to assure plumb stud alignment, the second top plate and ridge pole should be laid out together to assure rafter alignment. This holds true for all roof framing, although on valley and hip rafters the 16- or 24-inch module must be found against sloping, angled timbers.

Ridge Pole Placement

The ridge pole runs the length of the roof along a center line where the rafters from each side wall meet (Figure 137). Its long side is placed vertically. Because rafters that butt against it are cut on an angle, if the rafters are 2 × 8's, for example, the actual bearing surface against the ridge may be 8½ or 9 inches along the angled cuts. Consequently, it is customary to make the ridge pole from a dimensional timber one stock size deeper than the rafters, in this case, a 2 × 10.

There are several ways to place the ridge pole and rafters, but the surest way to avoid inaccurate cuts is to leave final rafter cuts until their limits, that is, the wall plates and the ridge pole, have been accurately and firmly established. Establishing the rafter boundaries makes the construction more understandable as the rafters become similar pieces of wood to fit into a box.

To place the ridge, first find the midpoint of the roof on each endwall of the house. Then cut a 2 × 6 equal to the rise plus 4 feet. Then, leaving at least 2 feet below the top plate, plumb this 2 × 6

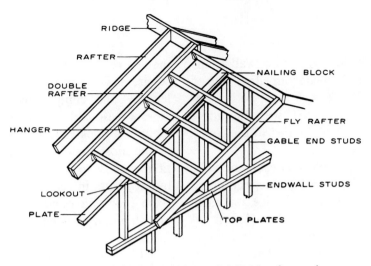

Figure 137. Typical ridge pole with parallel framing for overhang

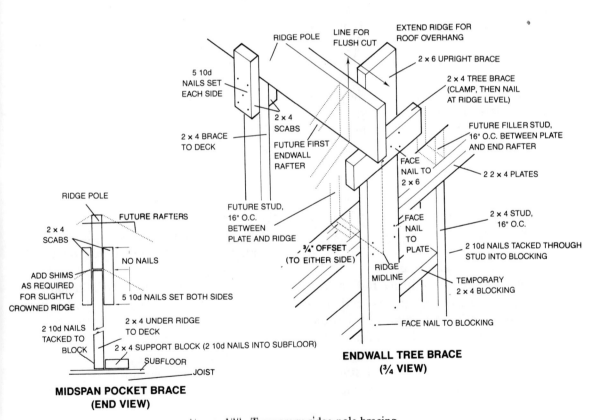

Figure 138. Temporary ridge-pole bracing

with the wide side against endwall framing for support, ¾ inch from either side of the midline. The 2 × 6 must be exactly vertical and well braced, with the addition of diagonal 2 × 4's tied to the framing if necessary.

With 2 × 6's in place at each endwall, measure the exact rise distance, and mark the bottom edge of the ridge on each 2 × 6 brace. Next use a short length of 2 × 4 to make a tree that can support the ridge in proper position. The 2 × 4 should extend about a foot to each side of the 2 × 6, be spiked to the 2 × 6 securely, and for extra strength, each end of the 2 × 4 (make sure it is level) may be braced down to the endwall.

Depending on the length of the house, several workers may be needed to lift the ridge pole into position on the braced trees. Place the crown up, and use a 4-foot level to verify the brace dimensions. If construction has been accurate up to this point, however, the equally dimensioned trees should produce a level ridge. Toenail the ridge to the braces, and add long 2 × 4 braces from midspan to the deck.

Rafter Cuts

The operation of cutting rafters to lie squarely against top plates and ridge poles is the crux of a solid roof. Up until now, framing joints have been square. But rafter cuts, which are just as critical as cuts on girders, joists, and studs, are angled and generally more complicated.

One somewhat technical method of determining rafter cuts is to use the mathematical relationships built into a framing square. The complete procedure is lengthy and complicated (Stanley provides a reasonable explanation in a manufacturer's booklet) and goes against one of the most basic principles of construction, that is, to minimize the use of new initial points in repetitive measurements. Each time you move the square and establish a new starting point there is room for error.

While roof pitch ratios can be held on the square to make bird's-mouth cuts accurately (the cuts where rafters are notched to bear across the top plate), it is risky to use them in a stepping-off process to determine the exact location of the ridge cut.

The point of a bird's-mouth cut is to provide a level seat for the rafter when it is set in an angled position (Figure 139). If no notch were cut, only hairline contact would be made between plate and rafter. So there are two lines to establish. The first is a level line, that is, it will be level when the rafter is erected. This is the plane that will sit on top of the wall plate. The second line is a plumb line, again, a line that will be plumb when the rafter is in place.

These two lines are perpendicular to each other, forming a right angle. Picture a stud wall, an end view, with the, say, 4 in 12 slope rafter rising to the right, and the overhang extending past the stud wall to the left. Now concentrate on the joint where the rafter and top plates meet. The level line on the rafter must be parallel to the top surface of the plate. The plumb

Roof Framing

line on the rafter must be parallel to the wall studs.

To find the level line lay the 4-inch mark on the short leg of a framing square, and the 12-inch mark on the long leg flush with the bottom edge of the rafter, with the point of the square (the corner where the two legs come together), down toward the ground and the long leg of the square toward the downhill end of the rafter. Note that this work is easily done with the rafter on the deck. However, when working on it, visualizing the timber in place is a help. Also available are small clamps, called buttons, that can be attached to the square to hold the slope ratio marks.

With the square in place, mark a line along the outside edge of the long leg. This is the level line that rests on the plate. Next, measure from the bottom edge of the rafter a 3½-inch distance along this line. This line segment represents the top of the bird's mouth, which is full wall depth. Now slide the framing square (maintaining the 4 and 12 marks flush with the bottom of the rafter) toward the rafter tail, or in your mind's eye, outside the wall. Align the outside edge of the short leg with the 3½-inch point on the level line, and draw a second line through this point toward the bottom of the rafter. This is the plumb line that will be perpendicular to the level line, forming a right angle at the back of the bird's mouth.

This procedure can be followed no matter what the rafter slope, so long as it is represented in mathematical miniature on the framing square. It is crucial, however, that sufficient material is left below the bird's mouth to form an overhang, if called for in the design, with an additional allowance for trimming the rafter tail.

This operation is further complicated by a process called stepping off—marking the rafter at 4 and 12 proportions again and again until the ridge limit is reached. It is the epitome of risking measurement error by working from new initial points with each repetitive dimension.

If the bird's mouth is accurately cut there is one relatively easy and risk-free method of determining the exact rafter length and angle of cut against the ridge. Mathematical proportions do not allow for field imperfections, but measuring and marking in place does.

To determine the ridge cut, place the bird's-mouth cutout on the top plate at the end of the stud wall. Then raise the high end of the rafter so that its top edge meets the top edge of the ridge pole. Hold or clamp the rafter in this position, and using the face of the ridge pole as a guide, scribe a mark where the rafter and ridge meet. This line should be parallel to the plumb line of the bird's mouth, and may be checked with the slope settings on the framing square. If there is a slight discrepancy, rely on the in-place mark. It represents reality, not the theoretically perfect outcome of a mathematical proportion.

When hip rafters or jack rafters running between wall plate and valley rafter are cut, it is essential to make a template. Scrap lumber of equal thickness to the rafter may be used, or simply a scrap of plywood trimmed to the actual, not nominal, rafter depth. Compound cuts must be tested on full thickness rafter material.

When one rafter has been cut, fitted in place, rechecked, and refitted, and you are satisfied that joints are tight and correct, mark "TEMPLATE" across both faces of the timber and use it as a pattern piece for all other common rafters, that is, those that run between plate and ridge.

Remember to crown each timber, and mark the crown edge with an X. This should be done on the template as well. When you set up sawhorses or some other system for efficiently mass producing rafters, take a look at the edges to make sure that the crown mark on the template is complemented by an X on the uncut timber.

At this point, rafter tails should be left uncut. Again, some builders will make these trim cuts on the deck before the rafter is placed. This assumes an unrealistic degree of perfection. And since the edge of the roof is one of the most visible and prominent straight lines on a house, easily recognized as true or not by eye, it can be trimmed most accurately after all rafters are placed.

RAFTER PLACEMENT

Although two workers may be needed to raise each rafter, there are simple steps that can be taken to facilitate placement and nailing. You should avoid unsafe conditions brought about by holding the rafter in place against the ridge with one arm while nailing with the other.

The bird's-mouth notch will hold the rafter on the wall plate. To hold the rafter against the ridge temporarily, cut strips of plywood or use 1 × 2-inch furring to

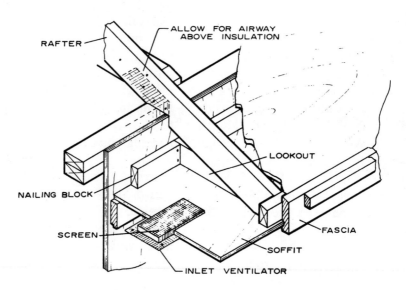

Figure 139. Rafter bird's mouth cut secured to stud wall plate

make a scab at the high end of the rafter. Nail strips roughly a foot long with 6 inches on the top edge of the rafter, and 6 inches protruding in line with the slope. The free end of this scab can be laid on top of the ridge while the rafter is aligned with plumb layout marks on the face of the ridge pole.

Rafters should be set in pairs to equalize loads on the ridge pole as construction proceeds. If rafter cuts have been made accurately and uniformly, do not recut them if discrepancies develop. These are most likely to be caused by lateral loads of the rafters already in place, pushing against the ridge and the outside walls. To protect against this the braces on perimeter walls should be left in place until all framing members have been installed.

The first rafter of each pair to be set may be end-nailed through the ridge pole. A 2 × 8-inch rafter would require three 16d end nails, and one 8d toenail through the top edge into the ridge. The second rafter in each set, however, must be toenailed; on a 2 × 8, a minimum of three

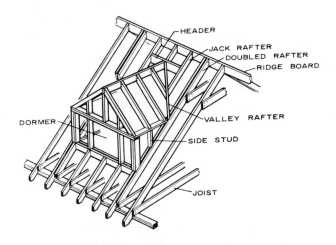

Figure 140. Dormer framing with valley rafters and compound cut jack rafters

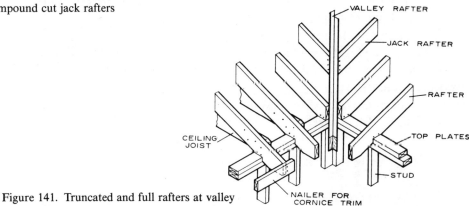

Figure 141. Truncated and full rafters at valley

8d nails from one side, and two from the other, should be used.

This nailing is not as insecure as it appears because the rafters tend to push against each other, through the ridge pole, as the load from roof decking and shingles increases. Still, it is wise to reinforce the toenailed rafter with a beam hanger, or a metal tie strap placed across the ridge, and roughly 12 inches down the edge of each rafter. These ties should be predrilled every 3 inches and secured with 10d box nails (thinner shanks than common nails to avoid the possibility of splitting).

Gable-End Studs

The roof line will create an open triangle between the endwall plates and the rafters. This space must be blocked in with 2 × 4 studding in line with the framing module in the wall below the plates (see Figure 134, page 207). Use a square to transfer stud locations from the bottom to the top of the plates, placing an X on the stud side of the square line. This joint poses no problems. Square cuts on the bottom of the filler studs are toenailed to the plate with two 8d or 10d nails from one side and one from the other.

The top joint, however, must bear against the angled rafter, which is single thickness (1½ inches wide), while the stud is 3½ inches. To make an extremely strong connection between the wall and the roof, use a 4-foot level to transfer the stud locations on the plate to the end rafters. Then cut 2 × 4's 5½ inches longer than the space between these plumb marks.

Set the stud in place with its narrow edge against the inside face of the rafter, and draw a line along the angle where the stud meets the rafter. It is helpful to set a bevel square to this angle, which will work (at different heights off the plate) for all filler studs. Now cut along this angled line 1½ inches into the 3½-inch stud depth, then start at the high end of the stud and make a rip cut (in line with the wood grain), 1½ inches from the outside edge, down to meet the first cut.

This detail leaves a 6-inch leg on the stud that can be face nailed to the inside of the rafter (Figure 143). Start the filler studs from the corners of the endwall, leaving the 2 × 6 tree brace for the ridge pole in place. As you approach the midpoint of the wall, a framed opening should be built to accommodate gable-end vents.

Determine the rough opening of the vent, and construct a header, supported by jack studs (see page 199). Cripple studs should be fitted to continue the framing module, and, if the layout does not work out to a midline stud, fit additional blocks directly under the ridge pole.

Rafter Tail Trimming

Determine the exact length of overhang on the trim rafters (the rafters above each endwall). Various tail designs can be used, bearing in mind that a fascia board and full-length gutter will be installed. Since blocking is added between rafter

Roof Framing

bays over the stud walls (see Figure 135, page 208), the rafter tails should not be subject to twisting. Still, it is good practice to cut the tails 1½ inches short of final overhang length to accommodate a structural fascia.

To establish a straight line, mark overhang length away from the stud wall on each trim rafter, set a nail at each mark, and connect them with a chalk line. Snap the line from the middle of the run, and transfer plumb cuts with a bevel square.

TRUSSES

Truss construction has been refined and improved in the last decade and is now used extensively by development builders. The truss is simply a triangle engineered to provide load strength over long spans without intermediate posts. Instead of gaining strength by increasing rafter width and depth, trusses are reinforced with web members connected to the triangle perimeter with plywood or metal gusset plates.

Typically, they are made of 2 × 4's, fabricated in the shop and transported to the site where a backhoe, bucket loader, or some other form of mechanical assistance is required to lift them into place. Clear spans of 32 feet and more are common, while conforming to strict deflection limits without support from partition walls. This increases floor plan options by eliminating interior bearing walls. In other words, floor space can be carved up according to use without regard to structural requirements.

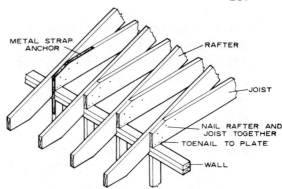

Figure 142. Rafter tail cuts for fascia and soffit applications

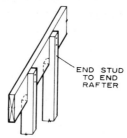

Figure 143. Gable endwall framing notched into last rafter

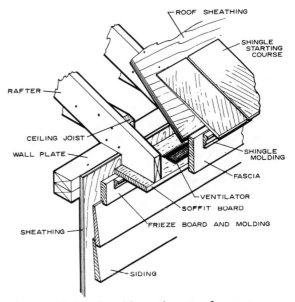

Figure 144. Details with venting at rafter overhang

The most common truss varieties are the W-type and the king-post type. A third configuration, the scissors truss, has a peaked bottom chord, and may be ordered to create a modified cathedral ceiling.

W-type trusses have 2 × 4-inch web members running between the midspan of both upper chords, to a point roughly dividing the lower chord into thirds, and from there up to the roof peak (Figure 146). A king-post truss has only a single web member running vertically between the midpoint of the lower chord and the roof peak (Figure 147). Its longer, unsupported upper chords limit spans more than W-type trusses, but it is extremely economical over moderate spans. For example, a plywood gusseted king-post truss with a 4 in 12 pitch, placed 2 feet on center, which is common for truss construction, is limited to approximately a 26-foot clear span (with 2 × 4 members). A similarly constructed W-type truss, however, could be used over a 32-foot span.

Trusses can be made at the job site, but this one-time assembly generally defeats the economy of high-speed shop assembly. Trusses are unwieldy, and built to withstand loads only in one direction. Consequently a large, flat area is needed

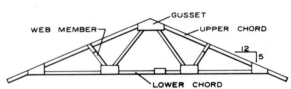

Figure 146. Typical framing of W-type roof truss

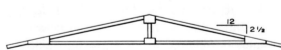

Figure 147. Simplified framing of king-post roof truss

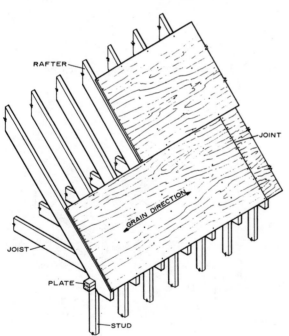

Figure 145. High slope rafters and sheathing

for assembly. Trusses should be stored and handled in a vertical position. Horizontal handling can weaken gusseted joints.

Trusses with horizontal bottom chords do have one distinct advantage over conventional framing. Chord assembly may be modified so that the triangle shape is cut short over the end walls. This box-end construction permits 12 inches of insulation in the ceiling and avoids the common loss of insulation efficiency when it is compressed over perimeter studwalls.

Throughout the structural phases of construction—building the footings, foundation, and frame—it is helpful to think of these elements as the skeleton of the house. When windows and doors are put in, roofing and siding are applied, and interior finishing materials are trimmed and painted, the skeleton will disappear. But its influence on durability, maintenance, and how much you enjoy living in your house will continue to be crucial.

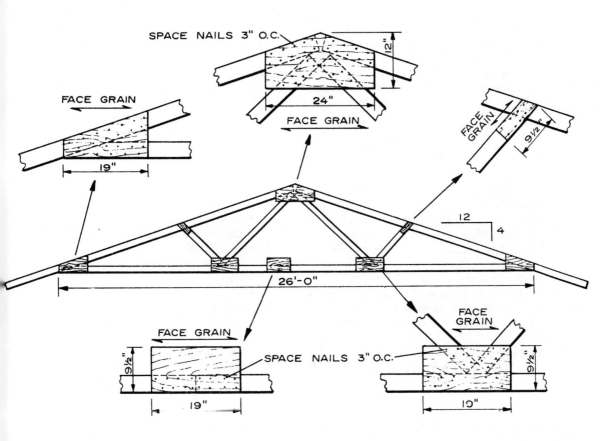

Figure 148. Gusset plate details for site-built roof trusses

FOUNDATION AND FRAMING INFORMATION SOURCES

American Concrete Institute, PO Box 19150, Redford Station, Detroit, MI 48219

American Hardboard Association, 205 West Touhy Ave., Park Ridge, IL 60068

American Institute of Timber Construction, 333 West Hampden Ave., Englewood, CO 80110

American Lumber Standards Committee, Suite 204, 20010 Century Blvd., Germantown, MD 20767

American Plywood Association, 1119 A St., PO Box 2277, Tacoma, WA 98401

American Society for Testing and Materials, 1916 Race St., Philadelphia, PA 19103

American Wood Preserver's Institute, 1651 Old Meadow Rd., McLean, VA 22101

California Redwood Association, 1 Lombard St., San Francisco, CA 94111

Forest Products Laboratory, Forest Service, U.S. Dept. of Agriculture, PO Box 5130, Madison, WI 53705

International Masonry Institute, 823 15th St. NW, Washington, DC 20005

Manufactured Housing Institute, 1745 Jefferson Highway, Arlington, VA 22202

National Association of Home Builders Research Foundation, Inc., 627 Southlawn Lane, Rockville, MD 20853

National Association of Home Builders of the United States, 15th and M Sts. NW, Washington, DC 20005

National Bureau of Standards, Center for Building Technology, Washington, DC 20234

National Forest Products Association, 1619 Massachusetts Ave. NW, Washington, DC 20036

National Ready-Mixed Concrete Association, 900 Spring St., Silver Spring, MD 20910

Northern Hardwood and Pine Manufacturer's Association Inc., 305 East Walnut St., Green Bay, WI 54301

Portland Cement Association, 5420 Old Orchard Rd., Skokie, IL 60077

Southern Forest Products Association, PO Box 52468, New Orleans, LA 70152

Western Wood Products Association, Yeon Building, Portland, OR 97204

U.S. Dept. of Housing and Urban Development, 451 7th St. NW, Washington, DC 20410

INDEX

Pages with illustrations are indicated by *italics*.

accelerators, 54
actual dimensions, 145
admixtures, 53–54
age-to-strength cement ratios, 59
aggregate, 53; effect on strength, 55
air-entraining, 53–54
air tools, 135–36
all-weather wood foundation, 91–92, *93*
American Concrete Institute, 53, 220
American Hardboard Association, 220
American Institute of Timber Construction, 220
American Lumber Standards Committee, 220
American Plywood Association, 220
American Society for Testing and Materials, 64, 220
American Wood Preserver's Institute, 220
anchor bolt, 59; at sill, 82, 193
antishorts, 170
appearance grades, 148
approved plans, 165
Architectural Graphic Standards, 27
average frost penetration, *46*
average weight of wood, 152

backer rod, 105
backfilling, 48, 100
backhoe, 45
back slope, 45
balloon framing, 179–81
bar clamps, 135
basement design, 43–44
batter boards, 40–43
bay extension, 196–97
beam hanger, 190–91
bearing edges: at floor, 190, 193, 203; at ridge, 210
bedrock foundation, 23, 46
bevel square, 127
bird's-mouth joint, 208, 212–13
bits, 133
bleed water, 108
block foundations, 63; construction details, 69–76; finishing, 79–82; layout, 67–68
blocking, 179–80
block plane, 133–34

board foot, 146
bond patterns, 66–67
box-end trusses, 218–19
box nails, 190
bracing: at batter boards, 43; at ridge, 212; at walls, 215
brickset, 37, *75*
bridging, 180, 186; on steel, 196
bright nails, 188
bubble vials, 127–28
building blueprints, 165
building codes: and construction, 165; and site selection, 27
building lines, 39–40; at corner stakes, 50
building sites: natural creation of, 11–12; man-made, 13; taxes on, 18–19
built-up block, *72*
built-up girders, 192–93
built-up roofing, 207
bull float, 108
buttering block, 70–71
buttons, 213

calcium chloride ($CaCl_2$), 54, 65
California Redwood Association, 220
cantilevers, 196
cant strip, 207
capping block, 80–82
carbide-tipped bits, 133
carpenter's level, 32, 34, 127
cast steel, 125; on circular saw, 129
cathedral ceiling, 205
cat's paw, 132
ceiling edge frame, 203–5
ceiling joist spans, 157, 158
cellar design, 166
cement mixes, 34
central-girder framing, 173
chalk line, 68; *69;* for carpentry, 128; on rafter tails, 217
chisels, 134
chord member, 205, 217
circular saw, 129–30
clamps, 134–35; at batter boards, 42
closure block, 73–75
collar beam, 205

combination blade, 130
combination square, 127
common nails, 187
common rafter, 214
compacted fill, 23
compound curing, 60
compound miter, 209
compressed-air tools, 135–36
compression: parallel to grain, 150–51; perpendicular to grain, 150–51
compression resistance, 5; of concrete, 88; of soil, 48
compression stress, 4–5
concrete block, 63–65; estimating data, 67
concrete consolidation, 90
concrete footings, 52–59
concrete forms, 82–88
concrete lifts, 90
concrete measurement, 53
concrete mixes, 54–55; for foundation, 88; and temperature variation, 90–91
concrete ordering, 55–56
concrete properties: for footings, 53; for foundations, 88; for slabs, 103–4
concrete volume conversion factor, 56
control joints, 79–80; at slab edge, 105
corner bracing, 174, 180; nailing for, 188
corner-cutting construction, xii–xiii
corner frame lapping, 198
corner posts, 200–202
corner stakes, 49–50
Coulomb, Charles Augustin de, 4
crawl-space design, 43–44; as plenum, 185–86
crimping tool, 169
cripple studs, 200–201
crossbridging, 174, 180
crosscut saw, 130–31
crowbars, 132
crowning, 192–93; of ridge pole, 212
crushing strength, 4
curing, 54; compound, 60; of concrete block, 64–65; strengths, 59–60
curvilinear construction, 113–15
cut and fill, 44–45
cut nails, 188
cutting edge tools, 124–25

dampproofing, *106*
darbys, 36; on slab finishing, 108
dead loads, 48, 152
deflection, 150–53, 154; of scaffolding, 136
design loads, 153–54
design values, 146–53
development building, 14, 16
dew point, 96–99
diagonal layout system, 40, *41*
dimensional timbers, 147

door frame, 180
door opening, 198–99
dormer framing, 215
doubled joists, 196
double-glazed fixed glass, 174
double-insulated tools, 129–30
drainage on site, 23–24
drain tile, 94, *95*
dressed timbers, 145
drills, 37, 132–33
drive screws, 169
drop-forged tools, 125–26; as hammers, 131
dry run layout, *69*
dry-wood termites, 99–100
dumpy level, 32–33
duration of load, 154

economic constraints, 188–89
edgers, 36
elasticity, 6–7
elastomeric caulk, 107
electric screwdriver, 169
end-grain nailing, 179, 187, 189, 215
end-matched planking, 182
end-wall framing, *178*
energy efficient orientation, 25
engineered construction, 171–72, *176*
environmental orientation, 25–27
Environmental Protection Agency, 100
equivalent thickness test, 64
excavating, 39; equipment for, 40, 45; volume of earth, 44
extension cords, 138–39
extension ladder, 136
extension ruler, 126–27
extreme fiber stress in bending, 149, 153–54

face nailing, 187, 189
face shell mortar, *74*
Federal Housing Administration, 98
Federal Trade Commission, homeowner report, 17
fiber saturation point, 145–46
fiber stress in bending, 149
fill, 23; for foundation, 46–47
filler block, 69, 71
fines (grit), 23
finished grade, 44, *45*
finishing trowel, 35
fire fighting, and site selection, 20–21
fireplace footings, 58
firestops, 179–80
first course block, *71*
fixed glass, 174
flathead bits, 133
flat roofs, 207
flat slope rafters, 159–61

Index

floating, 108
floats, 35–36
floor joist spans, 156
floor openings, *173,* 197
floor plan design, 25
footings, 9–10; construction sequence of, 38–39; design of, 48–52; layout of, 49; multilevel, 58–59
Forest Products Laboratory, 220
forged tools, 125–26
form design, 48–52; prefabricated, 87–88; site built, 83–87
form openings, 86
form reinforcement, 50–51; deflection of, 84
form release agents, 87
Forstner bits, 133
foundation control point, 44, *45*
foundation design, 43–44
foundation membrane, 95–96
foundation pockets, 193
foundation settling, 23
foundation ties, 84–86
frame components, 173; size estimates of, 142, *201*
frame construction, 191–202
frame design, 143–45
frame hardware, 188–91
framing alternatives, 165–70
framing costs, 177
framing square, 33; on rafters, 213
frost penetration, 46, *47*
frost resistance, 54
full mortar bedding, 70, *71*
furrowing mortar, 69–70, *71*

gable-end studs, 216
gable roof, 208
galvanized nails, 188
generators, 138
geodesic dome, 117–18
ghosting, 186
girders, 192–93
green timber, 145–46
ground clearance, 43
grounding, 138
groundwater sources, *12*
grout, 65; estimating data for, 68
gusset plate, 217–18, *219*

hacksaw, 131
hammers, 37; for carpentry, 131–32; holster for, 132
hand tool selection, 135
hardening, 124–25
hardpan soil, 17, 24
hawk, 35
header joist, 194, 197
headers, 199–201
high carbon steel, 124

high-slope rafters, 162–64
high-speed steel, 124
hip roof, 208–9
hoes, 35
homebuilding skills, xiv
home defects, 17
homeowner complaints, 17, 166
homeowner preferences, 25
home ownership duration, 19
horizontal shear, 150, 151
hot-dip galvanizing, 188
hydration, 59

I-beams, 191–92
imprint bleeding, 169
independent measurements, 127
induction motor, 139
initial points (IPs), 39
insulation: at foundation, 96–99; at slabs, 104–5
interlocking block, 79
International Masonry Institute, 220
interrupted joists, 196–97

jack plane, 134
jack rafter, 209
jack stud, 199–200
jamb block, *81*
job sequencing, 29–30
jointers, 36
jointing, at slab, 108
joint tooling, *76*
joist hangers, 190–91
joist openings, 196–97
joists, 194–95

keyed footings, 46, 51
king-post truss, 218
kneeboards, 36

labor costs, 177
ladder jacks, 136–37
ladders, 136
lag bolts, 188
lally columns, 191
laminated timbers, 183
land costs, 17
land development, 13
landlocked sites, 18
landscaping, in developments, 14, 16; complaints about, 17
lateral concrete pressure, 90
lean-to roof, 208
ledgers, 180; at rafters, 190
let-in bracing, *178,* 180; nailing for, 188
leveling, at batter boards, 42

levels, 30; carpentry, 127–28; site use of, 128; selection of, 32
lift placement, 90
lime, in mortar, 65, 66
linear construction systems, 113–17
linear foot, 146
line and blocks, 33–34; on corners, 73
line level, 33
lintels, 76, 77
live loads, 156–64
loadbearing wall, 167–68
load capacity, 149; by wood species, 151; by comparing bending and deflection, 153
load duration, 149, 154
longitudinal shrinkage, 152
lot development, 16; taxes on, 18–19
low-carbon steel, 124
low-slope rafters, 159–61
lugged footing, 46, *51*
lumber consumption, 165–66
lumber design values, 146–53
lumber grading, 148
lumber sizes, 145–47
lumber strengths, 142–45

machined tools, 125–26
maintenance planning, 19
major module, 172, 181
mansard roof, 208
masonry bits, 133
masonry properties, 28–30
masonry saw, 75
mason's level, 32
mason's trowel, 35
measuring, 31; by margin comparison, 33–34
medium-carbon steel, 124
medium-slope rafter, 162–64
membrane curing, 60
metal lath, 80, *81*
Michigan joint, 80
minor module, 172
mix calculation, 55–56
mix consolidation, 90–91
mix segregation, 56; at foundation, 90–91
modular construction, 30; at foundation, 62
modular design, 171–77; at windows and doors,174
modular dimensioning, 171–72
modulus of elasticity, 150, 151–53
moisture content, 145–46; when green, 152
mortar, 65–66; estimating data for, 68; for block, 69–70; removing burrs on, 76
Mylar tapes, 31

nailers, 203–5
nailing schedule, 189
nail pullers, 132
nails, 187
National Association of Home Builders, 220
National Bureau of Standards, 220
National Design Specification, 153, 220
National Forest Products Association, 220
National Ready-Mixed Concrete Association, 177, 220
natural site, 13
neighborhood selection, 19–21; demographics and, 22
new construction advantages, xii
nominal dimensions, 145
Northern Hardwood and Pine Manufacturing Association, 220

offset layout, 175–77
on center layout, 172
openings, 199–201
Orangeburg pipe, 94
orientation, 24–27
overbuilding, 141–43, 165
overhangs, *196*

paint brushes, 123–24
partition frames, 168
passive solar planning, 26–27
pedestal pier, 191
pennyweights, 187
perimeter belt, 188, 194
perimeter excavation, 40, *41*
pesticides, 100
piers, 58, 191
pilasters, 193
pile footings, 57–58
pilot holes, 188
pipe clamps, 135
planes, 133–34
plank and beam framing, 181–83
plank and beam rafters, 206–7
plates, *173*, 198, *199*
platform framing, 177–80
Plen-Wood framing, 184–85
plumb bob, 34, 68
plumbing (alignment): at foundation, 69, 71; at joists, 194; at rafters, 212–13; at studwall, 198
plywood forms, 83–84; on foundation, 91–92
portable mixers, 66
Portland cement, 53; in mortar, 65–66
Portland Cement Association, 220
posts, 191
poured foundation, 62–63
precuts, 198
prefabricated forms, 62, 87–88
pry bars, 132
pump jacks, 137
purchase price ratio, 17

Index

quenching, 124–25

Radburn superblock, 24–25
radial arm saws, 139–40
radial shrinkage, 152
rafter cuts, 212–14
rafter layout, 210
rafter placement, 214–19
rafter specifications, 154–55; and slope, 209–10, and slope conversion, 155, 210; and spans, 159–64
rafter square, 127
rafter tails, 216–17
rafter trusses, 217–19
rakers, 36
ready-mix concrete, 52–53
real estate tax, 18–19
rebars, 51–52; in foundations, 88; at openings, *90*
reinforcing wire, 104
relative humidity, 97–98
repetitive F_b, 149, 151
retarders, 54
retempering, 66
ribbon perimeter, 188, 194
ridge pole, 210–12
rigid insulation, 183
ripping bar, 132
ripping cut, with circular saw, 130; with table and radial arm saws, 139–40
rise, 209
roof frame design, 204–5; and construction details, 209–19
roof trusses, 205, 217
rough opening, 199
rulers, 31; for carpentry, 126–27
running foot, 146
R-value, 98–99; for slabs, 104–5

sand, 53
sawhorses, 137–38
saws, 29–31; blades for, 129
scabbing, 205; at rafters, 215
scaffolding, 136–37
scissors truss, 218
screed boards, 103–4
screeding, 56
scribing, rafter to ridge cuts, 213
seasoned timber, 146
secondary footings, 56–59
secure sites, 16; services for, 20–21
settling, 23
sheathing layout, 175
shed roof, 208
shoe, *173*, 178
shovels, 35
shrinkage, 152; in wall frame, 179–80
sills, 193–94; at openings, 200–201

sill sealer, 193, *195*
single F_b, 149, 151
site access, 40
site analysis, 14, 22–24
site-built forms, *60–61*, 83–86
site improvement, 17–19
site orientation, 14–16, 24–27
site utilities, 18
sixteen-inch module, 175–77
slab on grade, 43–44; design of, 101–103
sledge hammers, 132
sleepers, 107, *108*
slip plate, 76, *77*
slope, of rafters, 156–64, 209–10
sloping sites, *58*
snow load, 149
social orientation, 24–25
soffit, at cantilever, 197
soil bearing pressure, 23; at foundation, 46–48
soil dehydration, 48
soil identification, 23; on backslope, 45
soil mechanics, 22–24; at foundation, 46–48
solar orientation, 25
sole plate, *173*, 178
solid bridging, 180, 196; at gable end, 216
Southern Forest Products Association, 220
span limits, 153–54
span tables, 155–64
spreaders, 50, 86; for keyed footings, 51
spring clamps, 134
squares, 33, 127
squaring up, 41–42
standard penetration test (SPT), 23
star drill, 37
steel framing, 167–70, 186; gauge of, 168
steel hardening, 124–25
steel reinforcement, 51–52
steel tapes, 31, 126–27
stepladder, 136
stepped foundation, 23, 58–59
stepping ahead layout, 175–76
stiffness, 6–7
stone chisel, 37
story pole, 32–33; at block wall, 73
strap hanger, 196
stretchers, 63
stringer joist, 194
structural decking, 182
structural diagnosis, xi–xii
structural failure, 6; of foundation, 48
studwalls, 173, 178–80, 197–98; openings in, 199–200
stump removal, 17–18
subcontracts, 29, 45
subflooring, 177
subterranean termites, 99–100

sun angle, 26–27
surveys, 27

tablesaw, 139–40
taper-ground blades, 130
tempering, 124–25
template rafter, 208, 213–14
tension stress, 4–6, 8–9
termites, 99–100
termite shield, 100, 193
thermal design, 25–27
thickened-edge slab, 44
throw-away tools, 123
thrust line, 14
tiebar, 78
ties, 84–86
tie straps, 190, 216
tie wire, 52, 84
"tight-house" construction, 116
tilting arbor saw, 139
timber size, 145–47
tip-up construction, 178–79, 198–200, *201*
toenailing, 179; nailing for, 187; at rafters, 215–16
tongue-and-groove planking, 181
tool cost, 122–23
tool durability, 123–24
tooling, 76, *77*
tool selection, 121–23
tool steel, 124–25
top plate, 178–80
topsoil, 24
torpedo level, 33
town master plan, 18
trade associations, 220
transit, 32–33
tree protection, 40
trench footing, 39, *49*
triangulation, at foundation, 33; at batter boards, 40–41
trimming allowance, 172
trowels, 35
trusses, 217–19
truss joists, 184
tungsten-carbide alloy, 124
twenty-four-inch module, 175–77
twist drill bits, 133
tying, 56

U-factor, 97–99
undermining, 17
Unified Soil Classification, 23
universal motor, 139
U. S. Army Corps of Engineers, 6–7
U. S. Dept. of Housing and Urban Development, 220
U. S. Geological Survey, 23

vapor barriers, 94–96; for slabs, 104
variance appeal, 39–40
vibrator, 90

wales, 84, *85*
wallboard layout, *204*
wall header, 199–200, *201*
wall openings, 199–200, *201*
waste allowance, for mortar, 67; for timber, 72
water level tool, 32, 127; at batter boards, 42
waterproofing, 94–95, *96*
water saw, 36–37
watershed drainage, *12*
water-to-cement ratios, 54
weak links, 203–5
weather exposure, 13; of foundation, 94–99
web members, 218–19
western framing, 177–80
Western Wood Products Association, 184, 220
welded wire fabric, 107–8; cutting of, 37
wheelbarrow, 34; for mixing, 55
wind load, 149
window opening, 198–200, *201*
windrow, 69–70
wire cutter, 37
wire nails, 187
wood-boring bits, 133
wooden extension rule, 126–27
wood foundations, 91–92, *93*
wood species, 151
wrenches, 135
W-type truss, 218

yard drainage, 17

Z-bar, 80, *81*
zoning, 20, 22